Brain Mechanisms in Problem Solving and Intelligence

A Lesion Survey of the Rat Brain

CRITICAL ISSUES IN NEUROPSYCHOLOGY

Series Editors

Antonio E. Puente
University of North Carolina, Wilmington

Cecil R. Reynolds
Texas A&M University

ASSESSMENT ISSUES IN CHILD NEUROPSYCHOLOGY
Edited by Michael G. Tramontana and Stephen R. Hooper

BRAIN MECHANISMS IN PROBLEM SOLVING AND INTELLIGENCE: A Lesion Survey of the Rat Brain
Robert Thompson, Francis M. Crinella, and Jen Yu

BRAIN ORGANIZATION OF LANGUAGE AND COGNITIVE PROCESSES
Edited by Alfredo Ardila and Feggy Ostrosky-Solis

HANDBOOK OF CLINICAL CHILD NEUROPSYCHOLOGY
Edited by Cecil R. Reynolds and Elaine Fletcher-Janzen

MEDICAL NEUROPSYCHOLOGY: The Impact of Disease on Behavior
Edited by Ralph E. Tarter, David H. Van Thiel, and Kathleen L. Edwards

NEUROPSYCHOLOGICAL FUNCTION AND BRAIN IMAGING
Edited by Erin D. Bigler, Ronald A. Yeo, and Eric Turkheimer

NEUROPSYCHOLOGY, NEUROPSYCHIATRY, AND BEHAVIORAL NEUROLOGY
Rhawn Joseph

RELIABILITY AND VALIDITY IN NEUROPSYCHOLOGICAL ASSESSMENT
Michael D. Franzen

A Continuation Order Plan is available for this series. A continuation order will bring delivery of each new volume immediately upon publication. Volumes are billed only upon actual shipment. For further information please contact the publisher.

Brain Mechanisms in Problem Solving and Intelligence

A Lesion Survey of the Rat Brain

Robert Thompson

Late of University of California Irvine
Irvine, California

Francis M. Crinella

State Developmental Research Institutes
Costa Mesa, California

and

Jen Yu

University of California Irvine
Irvine, California

Springer Science+Business Media, LLC

Library of Congress Cataloging-in-Publication Data

Thompson, Robert, 1927-
Brain mechanisms in problem solving and intelligence : a lesion survey of the rat brain / Robert Thompson, Francis M. Crinella, and Jen Yu.
p. cm. -- (Critical issues in neuropsychology)
Includes bibliographical references p.
ISBN 978-1-4757-9550-9 ISBN 978-1-4757-9548-6 (eBook)
DOI 10.1007/978-1-4757-9548-6
1. Learning--Physiological aspects. 2. Brain--Localization of functions. 3. Rats--Physiology. 4. Neuroanatomy. I. Crinella, Francis M. II. Yu, Jen. III. Title. IV. Series.
QP408.T48 1990
599'.0188--dc20 89-26560
CIP

Originally published by Plenum Press, New York in 1990
Softcover reprint of the hardcover 1st edition 1990

To our wives: Pitsa, Terrie, and Janet

Whereas man's successes, persistences and socially unacceptable divagations—that is, his intelligences, his motivation and his instabilities—are all ultimately shaped and materialized by specific cultures, it is still true that most of the formal underlying laws of intelligence, motivation and instability can still be studied in rats as well as, and more easily than, in man.

—Edward Chace Tolman

It is easier to say where than it is to explain how. But localization of function in the brain must come first. . . .

—Wilder Penfield

Preface

This book is the outcome of a decade of research on the neuroanatomical mechanisms of learning in the young laboratory rat. It is essentially a discourse on the functional organization of the brain in relation to problem-solving ability and intelligence.

During the period between 1980 and 1989, well over 1000 weanling albino rats were subjected to localized brain damage (or sham operations in the case of the controls) under deep anesthesia and aseptic surgical conditions, were allowed to recover, and subsequently were tested on a wide variety of problems designed to measure general learning ability. Since virtually every part of the brain rostral to the medulla has been explored with lesions, it has become possible not only to map a number of "putative" brain systems underlying the acquisition of distinctive problem-solving tasks, but to isolate several neuroanatomical mechanisms that appear to be selectively involved in the acquisition of particular kinds of goal-directed learned activities. Of particular interest was the discovery of a "nonspecific mechanism" (previously referred to in our research reports as the "general learning system") inhabiting the interior parts of the brain.

One objective of this volume was to make these maps available in a single source. Another was to provide a description of learning syndromes arising from local lesions to different parts of the brain. A third objective was to determine, by means of correlational and factor analyses, whether the observed individual differences in problem-solving ability reflect variations in a "*g*" factor, or general intelligence. The final objective was to erect a conceptual framework within which these data could be interpreted. The resultant conceptualization, which is largely based on Penfield's "centrencephalic" theory, bears on several experimental and clinical issues related to the functions of the brain underlying higher mental activities. Some of these issues include the structure of animal (and human) intelligence, the neurology of learning, the distinction between cortical and subcortical dementia, and the neuroanatomical basis of mental retardation.

This book should be of use to those who are engaged in neurobehavioral research. It will be of particular interest to the increasing number of investigators who are searching for neurophysiological and neurochemical correlates of learning and memory or correspondences between the recently described chemical (and anatomical) pathways of the brain and behavior. Students and instructors in physiological psychology or behavioral neuroscience courses may also find the book informative, especially when a question arises concerning the effects upon learning of a lesion to a brain structure that is not given detailed discussion in whatever textbook they are using. Finally, it is certain to appeal to most readers having an interest in cognitive psychology or tending to take an extreme position in connection with the modular (localization) vs. equipotential (nonlocalization) models of brain organization.

As expected, an enterprise of this magnitude could not have been accomplished without the dedicated assistance of research associates. We are especially indebted to those workers who spent at least one year in our laboratory. Among these (listed in chronological order) are David Harmon, Kathy Gallardo, Peter Huestis, Victor Bjelajac, and Susan Fukui.

We also wish to pay tribute to our statistician, Todd Fisher, and to Phyllis Wood, who typed multiple versions of parts of this monograph.

Robert Thompson
Francis M. Crinella
Jen Yu

Contents

1

Introduction

This book deals with a review of some relatively recent research conducted in our laboratory that bears on the identification of those anatomical systems of the brain that might be concerned with learning. More specifically, it focuses on findings that help to delineate groups of neural structures involved in the acquisition of problem-solving tasks.

To write a comprehensive review of research related to this issue would be an awesome undertaking. Literally thousands of articles—ranging from lesion, drug, and electrical stimulation studies to investigation of the morphological, physiological, and chemical correlates of neuronal plasticity—constitute sources of information that are potentially useful in identifying the brain systems underlying learning. Even if such an attempt were made, the contradictions, confusion, and abundance of theory surrounding the literature on brain-learning relationships would probably prohibit any kind of synthesis. This state of confusion has developed, at least in part, because there are no standard learning tests within the experimental animal laboratory. Thus, some investigators have concentrated on habituation phenomena, discrimination learning, maze performance, or passive avoidance, whereas others have focused on classical conditioning, bar pressing, delayed response, or active avoidance. There is no reason to believe that the different laboratory situations invoke the same neural mechanisms of learning or produce associations having the same locus and distribution within the brain.

Confusion surrounding the literature on brain–learning relationships can also be linked to the difficulties involved in harmonizing the results obtained by one research method with those obtained by another research method. As a consequence, the different research technologies have generated somewhat independent stores of information from which different speculations about the neural circuitry of learning have arisen. One striking example of this concerns the question of whether discrete neural pathways are involved in learning and memory. According to lesion studies published within the last few years (Mishkin & Appenzeller, 1987; Thompson & Yu, 1987; R. F. Thompson, 1987), the answer to this question would be in the affirmative. However, the opposite answer is suggested by the study of electrical recordings of the activity of different brain sites during the execution of a learned response (John, 1967, 1972).

Probable interspecies differences in brain systems underlying learning may likewise contribute in no small measure to the contradictory literature. Consider the decorticate human who shows little evidence of learning (Deiker & Bruno, 1976), while the decorticate rat is capable of learning an assortment of laboratory problems (Oakley, 1983a).

This is not to say that the contradictions arising from the use of different behavioral tasks or research methodologies cannot be resolved or that similarities in the neurocircuitry of learning across the mammalian series are nonexistent. Clearly, there are numerous instances of convergence in the history of the behavioral neurosciences to warrant a diversified assault on the mysteries surrounding the neuroanatomy of learning. But learning in general and problem-solving in particular are extraordinarily complex phenomena that typically engage sensory, attentional, motivational, and motor processes besides associative and other cognitive mechanisms. Because of this bewildering degree of complexity, integration of knowledge within this domain is not expected to come easily.

In our view, one way to achieve some degree of insight into

the neurocircuitry of learning is to map the brain for structures critical for the acquisition of a given laboratory task. This can be accomplished by placing lesions in many different cortical and subcortical areas (using, of course, different groups of animals for different lesion placements) and then testing the animal's ability to learn. The distribution of lesions found to induce a significant learning impairment (relative to the performance of a sham-operated control group) would constitute a "lesion-defined functional system" governing the acquisition of the task under investigation. From this map of critical brain sites, correspondences to known neuroanatomical pathways could be inferred.

Obviously, the functional significance of each pathway represented within such a map cannot be determined from these data alone. However, inferences about the functional significance of each pathway could be developed if maps of additional laboratory tasks were charted. For instance, one particular pathway may be critical for the acquisition of a maze habit, but not critical for the acquisition of a visual discrimination habit. This finding would suggest that some distinguishing feature of the maze habit causes (or is correlated with) the emergence of this pathway in learning. This distinguishing feature might involve the necessity to construct a cognitive map or to rely heavily on the operation of working memory. To take another example of equal importance, a particular pathway may be found to be critical for the acquisition of a wide variety of tasks. In this case, the pathway in question would appear to have a nonspecific function, possibly playing a role in selective attention or in motor control.

For a number of years, we have been mapping those sites within the rat brain critical for acquisition of a variety of laboratory tasks. To date, four distinct habits have been mapped, including a visual discrimination, vestibular–proprioceptive–kinesthetic (inclined plane) discrimination, three-cul maze, and a "climbing platform" detour problem. (Actually, a series of three climbing detour problems was studied, but only the data derived from the initial platform problem have been mapped because they are uncontaminated by "transfer effects" arising from learning of previous detour problems.) A partial map of brain sites concerned with acquisition of puzzle-box/latch-box problems has also been constructed.

Choosing the lesion method, on the one hand, and laboratory rats, on the other, to study the neuroanatomy of learning was dictated by several considerations. First, lesion studies still constitute one of the most reliable and straightforward approaches to acquire clues concerning the identity of those nuclei and pathways implicated in learning. To paraphrase a remark made by one neuroscientist (Iversen, 1973), "without lesions, there would be very little hope of understanding the neurological circuitry of learning." Second, the rat has been one of the most popular laboratory animals used in lesion studies on learning. In fact, virtually every part of the brain of this animal has been subjected to investigation by the lesion method. Finally, probably more is known about the cytoarchitecture, chemoarchitecture, and connectivity of the different regions of the rat brain than of any other laboratory animal (Paxinos, 1985a,b).

Young albino rats (about 3 weeks old at the time of surgery and approximately 6 weeks old when introduced to the first learning task) rather than adults were investigated in this series of experiments on the effects of brain damage on learning. Choosing the young rat for study was primarily based on developing a brain-injured animal model of mental retardation (Thompson, Huestis, Crinella, & Yu, 1986), an objective that required the use of immature subjects. While the resultant data were not always found to be the same as those obtained with the use of adult subjects, there were clearly more similarities than differences (see, for example, Thompson & Yu, 1987, and compare the neural mechanisms described in Chapter 5 with those reported in adult rats in Thompson, 1982a).

Learning vs. Problem Solving

It is important to note that the tasks that have been chosen to undergo analysis by the lesion method are not representative of all types of learning. Learning, which is often defined as a relatively permanent change in behavior as a result of experience, is not a unitary phenomenon. There may be nonassociative learning, stimulus–response learning, cognitive map learning, taste aversion learning, semantic learning, learning to learn, and the like. The dichotomy we wish to emphasize is learning in a problem-solving situation vs. learning in a nonproblem-solving situation. With respect to the former, there is a goal (reward) to achieve. It may be water in the case of a thirsty subject or an area of safety in the case of a subject encountering noxious stimuli. Furthermore, the subject is required at some point within the situation to make a decision based on previous experience either to respond this way (approach this route), that way (climb that platform), or not at all (refrain from pressing a lever) in order to gain access to the goal. Laboratory tasks that would be regarded as invoking problem-solving processes include the discrimination, maze, detour, and puzzle-box problems that have already been singled out for investigation. Laboratory tasks that are not envisaged as initiating problem-solving processes include habituation, sensitization, classical conditioned responses, experimental extinction, passive avoidance responses, taste aversions, and latent learning. Parenthetically, if learning in general is defined as "the acquisition of knowledge through experience" (Revusky, 1985), then problem-solving would involve the utilization of learning for the purpose of attaining a goal.

This book deals mainly with the neuroanatomy of learning in problem-solving settings, and unless specified otherwise, the use of the word learning refers only to this class of problems.

Defense of the Lesion Method

This section serves two purposes. As the subheading implies, one purpose is to defend the use of the lesion method in the study of the neuroanatomy of learning. The second purpose is to highlight some of the complex problems associated with the investigation of brain–learning relationships. These problems are, in most cases, of such a general nature that they confront the investigator no matter what research technology is used.

Most readers are probably well acquainted with some of the weaknesses of the lesion method as it applies to the study of learning and memory (see Isaacson, 1976; John, 1972; Lynch, 1976; Shoenfeld & Hamilton, 1977; Webster, 1973). Since this monograph deals almost exclusively with empirical findings obtained by the lesion method, it is necessary to examine in detail some of the more common criticisms of this approach. Some of these criticisms are either exaggerated or ill conceived, and others are thoughtful and relevant but hardly serious enough to discredit the technique. For purposes of exposition, these criticisms will be expressed as propositions couched in terms of lesion-induced learning impairments. A similar defense of the lesion method was advanced earlier in connection with the study of brain–memory relationships (Thompson, 1983).

Some Criticisms

All lesion-induced learning deficits are due to disturbances in motor function, emotion, motivation, sensory–perceptual capacities, and the like rather than to destruction of nuclei or pathways directly concerned with learning.

This is by far the most common charge against the lesion method as it applies to learning. Clearly, there are numerous instances in which a learning deficit is due to disturbances in performance rather than destruction of learning circuits. Perhaps the most spectacular demonstration of this concerns those studies reporting that a second (or third) lesion superimposed on an initial lesion can reduce, at least in part, a learning deficit brought about by the initial lesion (Fonberg, 1975; Irle, 1985; Irle & Markowitsch, 1983, 1984; Meyer, Johnson, & Vaughn, 1970; Stokes & Thompson, 1970; Taghzouti, Simon, Herve, *et al.*, 1988; Vanderwolf, 1964). But to make the claim that *all* lesion-induced learning impairments reflect performance disorders independent of any impairment in learning ability is clearly begging the question and simplifying a bewilderingly complex problem for interpretation. For example, it is quite possible that a lesion-induced learning impairment that is accompanied by an emotional–motivational disturbance is the result of the destruction of a part of the brain that plays a role in both associative and emotional–motivational processes. Those theories of behavior that state that learning consists, in part, of a connection between stimulus events and emotions would predict such a result (Konorski, 1967; Mowrer, 1960). A similar argument can be developed in the case of a lesion-induced learning impairment that is attended by a motor or sensory–perceptual disorder.

There is another argument that weakens the proposition under discussion. Many cases have been reported in which brain damage may lead to severe alterations in behavior *without impairing acquisition of certain problem-solving tasks*. Some specific illustrations are the following:

Motor Disturbance

Rats sustaining almost complete destruction of the cerebellum exhibit the classic syndrome of locomotor ataxia involving a staggering locomotor pattern during forward progression. Despite this motor handicap, such ataxic rats have been found to learn a black–white discrimination (involving a long approach response to the discriminanda) and three detour problems (involving climbing responses) about as fast as controls (see Chapter 3).

Emotional Disturbance

Damage to the medial portions of the anterior (or midtuberal) hypothalamus in the rat has been reported to produce a syndrome consisting of prolonged savageness, hypersensitivity to tactile and aversive stimuli, and deficits in acquisition of active and passive avoidance conditioning (Cardo, 1961; McNew & Thompson, 1966; Thompson, 1978a,b). However, rats with anteromedial hypothalamic lesions are not impaired in learning appetitively motivated detour problems (see Chapter 3).

Motivational Disturbance

If a motivational involvement coexists with defective learning ability in a brain-damaged subject, the investigator is expected to report this observation. Motivational alterations are among the easiest of all performance disturbances to detect—the subject simply does not respond with normal vigor in pursuit (or in the presence) of the goal object. Therefore, if a lesion-induced learning impairment is not accompanied by an observable disturbance in motivation, then attribution of the learning impairment to a suspected motivational involvement is hardly defensible.

Even the presence of a motivational disorder may not, under certain circumstances, be associated with poor learning scores. Young rats with lesions to the frontal cortex, red nucleus, or inferior colliculus are likely to exhibit a disturbance in the execution of escape–avoidance responses to foot shock in a Thompson–Bryant visual discrimination apparatus (they require more frequent foot shocks to force escape responding than the controls); yet these animals learn an aversively motivated white–black discrimination problem about as fast as controls (Thompson *et al.*, 1986; Thompson, Huestis, Crinella, & Yu, 1987).

The extent to which any given lesion will produce impairments in learning is determined by the magnitude of the lesion rather than its locus.

This proposition, of course, is derived from Lashley's (1929) principle of "mass action." While this concept has been found to be wanting in the light of subsequent observations on both brain-damaged rats (Spiliotis & Thompson, 1973; Thomas,

1970; Thomas & Weir, 1975; Thompson, 1969; Thompson *et al.*, 1986) and humans (Gazzaniga & LeDoux, 1978; Piercy, 1969; Zangwill, 1964), it nevertheless has been invoked either as one line of evidence favoring nonlocalization of learning and memory processes or as another example of the impotency of the lesion method in identifying the neuroanatomical substrates of problem-solving activities.

There is little doubt that the mass action effect can roughly be demonstrated in certain maze situations, at least with respect to neocortical lesions in the rat. It is even possible that learning and retention deficits observed in certain situations involving conditioned avoidance behavior might be a function of the size of the lesion rather than its topography (Meyer *et al.*, 1970; Saavedra, Pinto-Hamuy, & Oberti, 1965). The important point is that the locus of the lesion rather than its magnitude is the critical factor in determining the presence of learning deficits on visual discrimination (Thompson *et al.*, 1986, 1987), vestibular–proprioceptive–kinesthetic discrimination (Thompson *et al.*, 1986, 1987), tactile discrimination (Simons, Puretz, & Finger, 1975; Zubek, 1951), detour (Thompson, Harmon, & Yu, 1984), and puzzle-box (Thompson, Bjelajac, Huestis, *et al.*, 1989a) problems.

In view of these data, the mass action effect must be considered to be idiocyncratic to but a fraction of the spectrum of laboratory tasks on which the brain-damaged rat has been examined. Consequently, the mass action effect cannot reasonably be offered as evidence for or against any position regarding either the diffuseness of the learning mechanisms within the brain or the utility of the lesion method to identify these mechanisms (see also Kolb & Whishaw, 1988).

Lesion-induced learning impairments are likely to arise from dysfunctioning of brain areas distant to the site of the lesion rather than from destruction of brain tissue at the site of the lesion.

It is becoming increasingly clear that morphological, physiological, and biochemical changes can take place in brain regions distant to the site of a lesion (Dauth, Gilman, Frey, *et al.*,

1985; Deuel & Collins, 1984; Lynch, 1976; Shoenfeld & Hamilton, 1977). These changes may be found in nerve cells that send their axons to the lesioned area as well as in nervous structures that receive afferents from the lesioned area. The implication of these findings is that a lesion, no matter how discrete, may disrupt normal functioning of brain circuitry left intact by the lesion and that this disruption may be responsible for any ensuing learning impairment.

This argument is certainly compelling and cannot readily be dismissed. However, like other criticisms of the method, it lacks precision in predicting the outcome of lesion experiments on learning. Consider, for example, the cerebellum, which has widespread connections with the spinal cord, medulla, pons, midbrain, and diencephalon (Voogd, Gerritts, & Marani, 1985). Damage to this structure might be expected to produce morphological, physiological, and biochemical changes throughout much of the brain. Yet, destruction of the cerebellum in the rat does not significantly retard learning of certain visual discrimination and detour problems (see Chapter 3).

The absence of a learning impairment following local brain damage is insufficient evidence for the conclusion that the area damaged is not concerned with learning.

In summarizing about 1100 connections within the rat brain, Knook (1965) chose not to construct functional maps for many reasons, one of which being that it would be "impossible to imagine which of the more than 17 anatomical possibilities a cortical impulse will use to reach the substantia nigra" (p. 435). This instance of anatomical redundancy within the central nervous system forms the basis for believing that negative findings associated with destruction of one pathway or nucleus within the brain do not constitute a solid base on which to infer that the damaged area is not a participant in the behavior under discussion. For example, two structures playing a critical role in learning may function in parallel so that the elimination of one would produce little or no effect on learning, whereas eliminating both would produce powerful effects on learning.

The essential point in connection with this proposition is that specific (and nonspecific) learning impairments do occur as a result of a single bilateral lesion, despite the presence of redundancy. Depending on the nature of the task investigated, learning (or retention) deficits may follow focal lesions to virtually any part of the neocortex, basal ganglia, limbic system, thalamus, hypothalamus, midbrain, or pons (Thompson, 1978a; see Chapter 3).

Summary

The foregoing analysis of some of the criticisms of the lesion method was not intended to free the method from any or all objections ascribed to it. Rather, its purpose was to draw attention to the fact that the grounds on which the criticisms are based are themselves subject to criticism. Surely, the lesion method, like other methods used in neurobehavioral research, has its weaknesses, limitations, and pitfalls. The uncertainty involved in interpreting behavioral changes following local brain damage constitutes a major weakness of the method. Another is manifested in the fact that a typical aspirative or electrolytic lesion destroys both fiber pathways and nerve cells, thus making it difficult to determine whether the former or latter (or both) is responsible for any behavioral changes. In spite of these weaknesses, the historical fact remains that the lesion method has yielded findings that have contributed greatly to our understanding of brain function and therefore may still hold promise in uncovering the neurocircuitry of learning. At the very least, lesioning specific structures of the brain constitutes the method of choice in testing the validity of any given neurological theory of learning.

Plan of the Book

In Chapter 2, details about the animals, behavioral methods (tests), surgical and histological procedures, and the manner in which the results are mapped on brain sections are given.

Chapter 3 presents the illustrations that constitute the main body of data. At the beginning of the chapter, there is a list of brain structures where the lesions were placed. Each line drawing of a brain section showing the site of the lesion and its associated syndrome of learning deficits is accompanied by a photograph of a corresponding unstained brain section appropriately labeled and furnished with frontal, lateral, and vertical stereotaxic coordinates, scaled in millimeters. These illustrations are organized in a manner similar to that of a stereotaxic atlas. The first two diagrams present the cerebral surface along with the sites of lesions to the cerebral surface, olfactory bulbs, and cerebellum. These are followed by a diagram of a parasagittal section of the brain showing the locus and extent of damage to two separate regions of the cingulate (limbic) cortex. In subsequent illustrations, diagrams of frontal sections through the brainstem are used to depict the sites of lesions to the various structures located within the interior parts of the brain. These frontal sections proceed from anterior to posterior levels.

It is possible to cast the lesion data presented in Chapter 3 into maps of "learning systems." This can be done by plotting the distribution of lesions that produce a learning deficit on a particular task. Since four different problem-solving tasks were investigated, four maps would result. At the beginning of Chapter 4, there is a list of the learning systems along with the number of the figure (a parasagittal section) that provides the corresponding map. The maps referring to specific learning systems follow. These maps constitute a unique source of clues concerning the neurocircuitry underlying problem-solving behaviors.

Based on a comparison of the maps of critical structures associated with each of the four problem-solving tasks, it has been possible to identify a nonspecific mechanism—a group of brain structures involved in the acquisition of all four tasks (as well as the acquisition of two additional climbing detour problems and a series of puzzle-box problems). It has also been possible to isolate a number of neuroanatomical mechanisms that appear to be selectively involved in the acquisition of particular kinds of goal-directed learned activities. Chapter 5 is devoted to a discussion of these specific and nonspecific neuroanatomical mechanisms.

Chapter 6 (largely written by F.M.C.) is unique in that it focuses on the study of individual differences in problem-solving ability among the control and brain-damaged rats. With the use of various correlational and factor analyses, an attempt was made to determine if performance on the different learning tasks is positively correlated and if it can be assumed that a general intelligence factor, or "*g*," exists at the level of the laboratory rat. Additional psychometric methods were applied to ascertain the extent to which specific factors account for the data and the degree to which individual brain structures are associated with general and specific factors. Interestingly, emerging out of this series of analyses was a distinction between psychometric *g* and biological *g*.

Chapter 7 deals with a discussion of the foregoing data and the introduction of a conceptual framework (based on Penfield's "centrencephalic theory") within which these data can be interpreted. As shown in Chapter 8, this conceptual orientation has implications bearing not only on the neurocircuitry of learning, but on such issues as the neurology of intelligence, the neuropathology of subcortical dementia, and the neuroanatomy of mental retardation. Finally, Chapter 9 details the ease with which centrencephalic theory can now be tested, despite the fact that many predictions derived from this theory coincide with those derived from more corticocentric theories.

2

General Methods

Subjects and Surgery

Weanling (21- to 25-day-old) male Sprague–Dawley albino rats, 50–65 g, underwent surgery under deep chloral hydrate anesthesia (400 mg/kg). Cortical lesions were accomplished by aspiration, while all subcortical lesions were made stereotaxically by passing a constant anodal current of 1.0–2.5 mA for a duration of 5–10 sec through an implanted stainless steel electrode (0.5 mm in diameter) with 0.5–1.0 mm of the tip exposed. Sham-operated control animals underwent the same surgical procedure as the experimental animals, except for the drilling of the skull and the insertion of the suction tip or lesion electrode into the brain.

The specific surgical procedure was as follows: After the rat was anesthetized and the hair over the cranium closely shaved with electric clippers, its head was positioned (oriented horizontally) in the headholder. The cranium was painted with tincture of iodine after which a longitudinal incision was made with a scalpel, starting at eye level and proceeding posteriorly to include skin halfway down the neck. The tissues over the skull were reflected laterally with the handle end of the scalpel, and any blood from the exposed surface of the skull was removed with gauze sponges. If the cortex, olfactory bulb, or cerebellum was the target of ablation, small holes were drilled over the appropriate brain area and widened with rongeurs. The dura was cut with a needle, retracted with forceps, and a suction tip (#18 gauge hypodermic needle) connected to a portable aspirator by means of rubber tubing was used to remove the brain tissue. When subcortical lesions were desired, small holes were made in the appropriate area of the skull to accommodate the passage of the lesion electrode. With the use of a stereotaxic instrument, the lesion electrode was lowered into the brain. From the lesionmaker, the anodal lead was attached to the electrode, and the cathodal lead was clamped to the exposed muscle of the neck. The current was then applied. When all bleeding stopped, the skin was sutured with 9-mm wound clips, and povidone-iodine (10%) ointment was applied liberally to the surface of the wound.

Following the surgery, the animal was placed into a separate cage and covered with a towel to minimize heat loss during the period of anesthetization. The animals were subsequently housed, two or three per cage, in medium-size hanging wire cages containing a constant supply of food pellets and water. A dish of sweetened wet mash was placed daily in each cage during the first postoperative week to encourage early resumption of food intake. During the third postoperative week, the animals were handled daily for approximately 5 min. During this handling period and the subsequent period devoted to testing of problem-solving ability, the experimenters were unaware of the group to which each subject belonged. All animals were maintained on a 12-hr light–dark cycle with lights on at 0600 and were trained only during the light phase.

Apparatus

Appetitively Motivated Problems

Climbing Detour Problems

The detour apparatus, whose dimensions have been reported elsewhere (Thompson *et al.*, 1984a), was divided into a start box painted flat white, choice chamber painted flat white, and goal box painted flat black. An opaque guillotine door separated the start box from the choice chamber. Interchangeable partitions fitted with a platform, cylinder, or ladder (each of which was painted flat white) could be positioned between the choice chamber and goal box to form the three detour problems (Figure 2.1). During preliminary training, a partition containing a centrally located window at floor level was employed. For Problem A, the partition used in preliminary training was placed in the apparatus along with a platform that sloped upward into the choice chamber to a maximum height of 10.1 cm above the floor. Problem B consisted of a partition containing a centrally located plastic cylinder that extended 20.3 cm into the choice chamber and was elevated 5.7 cm above the floor. Problem C consisted of a partition containing a window located above the floor that could only be reached by climbing a vertically positioned ladder that extended 19.2 cm into the choice chamber. (For Problems B and C, a ramp located behind the partition allowed the rat to descend to the floor of the goal box.) The entire apparatus was covered by a transparent Plexiglas lid and was illuminated by conventional ceiling fluorescent lights.

Puzzle-Box Problems

For reasons to be mentioned later, these problems were given to only 11 brain-damaged groups (those with lesions to either the frontal cortex, parietal cortex, dorsal hippocampus, dorsal caudatoputamen, globus pallidus, ventrolateral thalamus, substantia nigra, ventral tegmental area, superior colliculus, median raphe, or pontine reticular formation) for the purpose of assessing their motor learning capacities. The apparatus consisted of a start box painted flat white, a choice chamber painted flat white, and a goal box painted flat black. A clear Lexan guillotine door separated the start box from the choice chamber. Interchangeable smoked Lexan partitions, each containing a Lexan door, could be interposed between the choice chamber and goal box. During preliminary training, a "conventional" door that could be pushed open was used. Partitions containing doors that were employed during experimental

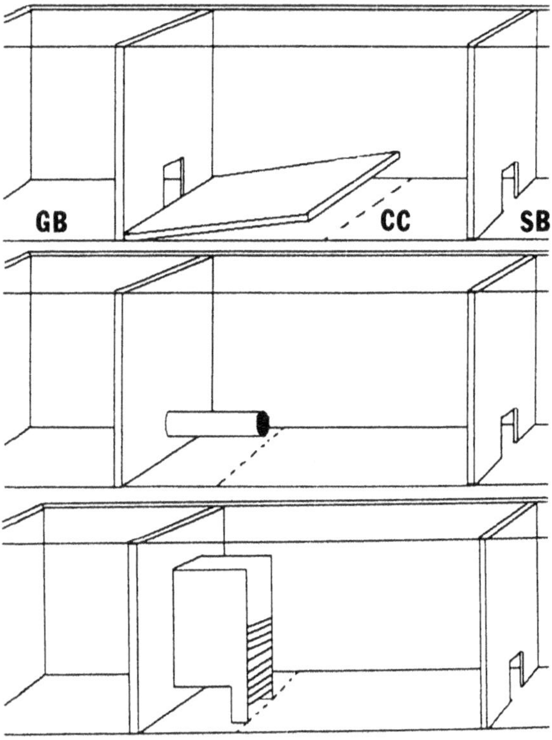

FIGURE 2.1. Schematic drawing of the detour apparatus showing the start box (SB), choice chamber (CC), and goal box (GB). The interrupted lines mark the boundaries of the "blind alleys." On Problem A (top panel), the rat must mount the raised platform in order to gain access to the goal box. On Problem B (middle panel), the rat must enter the elevated cylinder to reach the goal box. On Problem C (bottom panel), the rat must climb the ladder to reach the goal box.

training are shown in Figure 2.2. The entire apparatus was covered by a Plexiglas lid and illuminated by conventional ceiling fluorescent lighting.

Problems 1–4 contained doors with no latches. Access to the goal box could be achieved either by sliding the door to the left (Problem 1), pushing the door upward (Problem 2), pulling the door outward (Problem 3), or pulling the door outward and then downward (Problem 4).

Problems 5–8 contained doors that could be pushed open only after a latch was unlocked. The correct motor act consisted of either rotating the butterfly latch mounted on the choice chamber side in a clockwise direction (Problem 5), rotating the butterfly latch mounted on the goal box side in a clockwise direction (Problem 6), sliding the barrel bolt to the right (Problem 7), or elevating the hook (Problem 8).

Aversively Motivated Problems

Visual Discrimination

A two-choice Thompson–Bryant discrimination box, utilizing the motive of escape–avoidance of mild foot shock (1.0–1.5 mA), was employed (Figure 2.3). It consisted of a start box, choice chamber, and goal box. The floor of the start box and choice chamber was made into a grid, whereas the goal box floor was constructed of wood. Two windows, 14.0 cm^2, at the far end of the choice chamber provided the only means by which the animal could enter the goal box. A pair of gray cards mounted on wooden blocks was used in preliminary training. The stimuli for the visual discrimination problem consisted of a white card and a black card.

The entire apparatus was painted flat black, except the grid floor, vertically sliding clear Plexiglas start box door, and clear Plexiglas lid. Illumination of the apparatus was supplied by conventional ceiling fluorescent lights and a pair of fluorescent lamps (20 W each) placed 2.0 cm above the lid of the choice chamber.

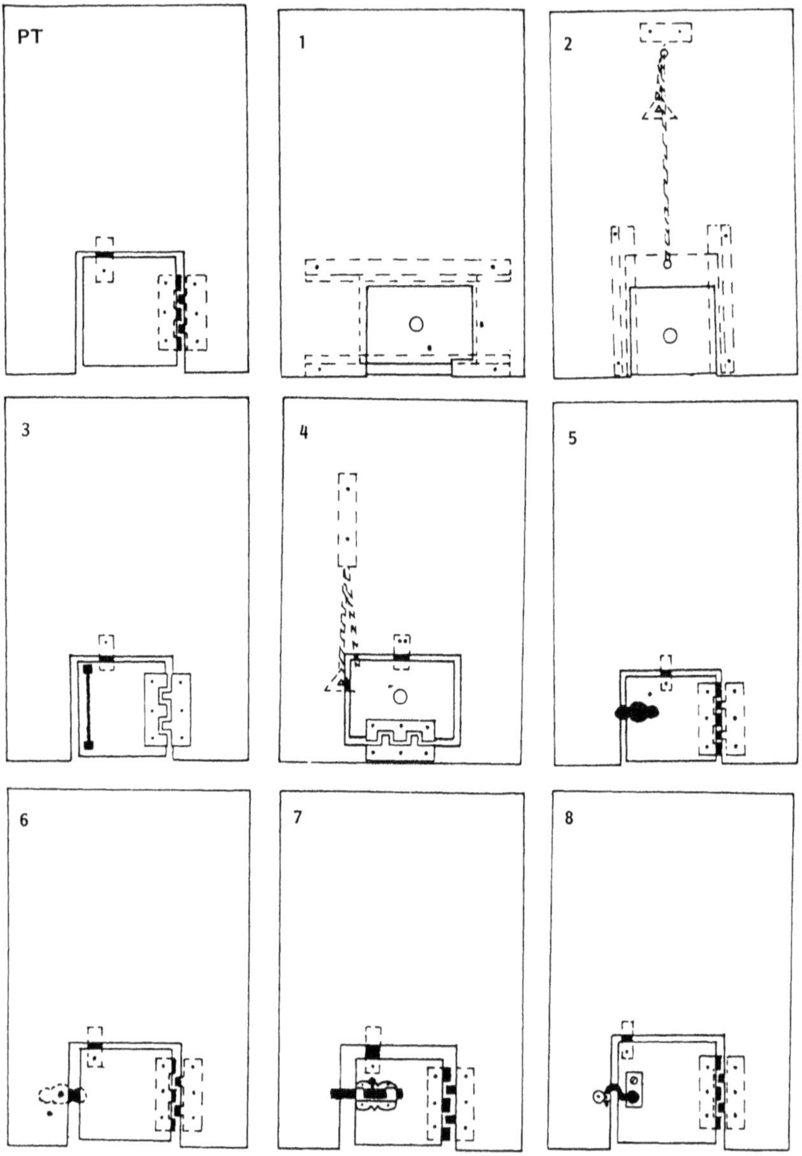

FIGURE 2.2. Front view of the interchangeable partitions containing doors showing the door used in preliminary training (PT) and the eight puzzle-box problems.

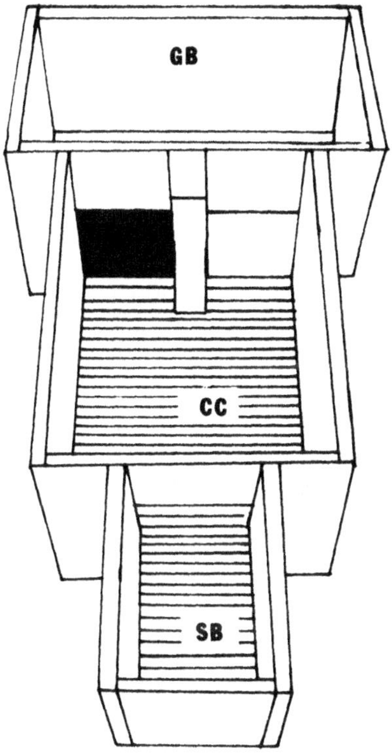

FIGURE 2.3. Schematic drawing of the visual discrimination apparatus showing the start box (SB), choice chamber (CC), and goal box (GB).

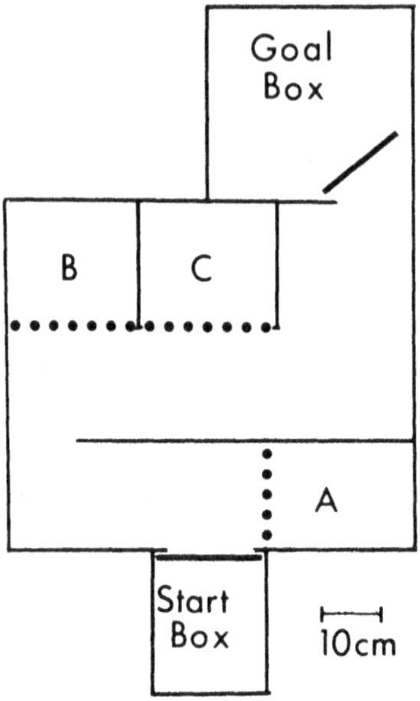

FIGURE 2.4. Floor plan of the maze showing the boundaries between the true path and each blind alley (heavy dotted lines) along with the start box, goal box, and goal box door (heavy solid line).

Maze

An enclosed three-cul maze was employed, which was designed for the use of the motive of escape–avoidance of mild foot shock (Figure 2.4). The start box and maze proper contained a grid floor, while the goal box floor was made of wood. The true path, which measured 120.5 cm from the start box exit to the goal box entrance, consisted of a 90° turn to the left (avoiding the first cul), a 180° turn to the right (avoiding the second and third culs), followed by a 90° turn to the left. The entire apparatus was painted flat black, except for the grid floor and clear Plexiglas lid. Illumination was provided by conventional ceiling fluorescent lights. No effort was made to control for any intramaze olfactory cues or extramaze cues.

Inclined Plane Discrimination

An enclosed single unit T maze, adapted for the use of the motive of escape–avoidance of mild foot shock, was employed to develop the vestibular–proprioceptive–kinesthetic discrimination habit (Figure 2.5). The stem of the T served as the start box and the left and right arms constituted the choice chamber. At the end of each arm was a window through which the rat could enter the end box by pushing aside a black card mounted on a wooden block. The stem and arms of the T had a grid floor; each end box floor was made of wood. The entire apparatus was secured to a platform that could be tilted 11° laterally and was illuminated by conventional ceiling fluorescent lights.

Procedure

Following a 3-week recovery period, the various groups of animals began training on the test battery. Some of the animals from each group were tested only on the climbing detour problems, others were tested only on the discrimination and maze problems, and still others were tested on the detour, discrimination, and maze problems. (As mentioned earlier, certain brain-damaged groups were tested only on the puzzle-box problems.) In the latter case, the appetitively motivated problems were learned first. The various groups of animals were not treated uniformly because of the evolving nature of this research project; however, a corresponding control group was invariably observed concurrently with any given brain-damaged group, permitting the performance scores of the latter to be evaluated statistically for the presence of any significant impairments in learning.

Appetitively Motivated Tasks

Climbing Detour Problems

After 2 days of water deprivation, each animal was allowed to explore the entire apparatus (Day 1). A dish of water as well as a dish of sweetened wet mash were available in the goal box, from which the animal was permitted to ingest for 10 min. On Day 2, the animals (usually run in squads of two or three) were given 10 preliminary training trials with an intertrial interval of 90–180 sec. Each trial began by inserting the animal into the start box and raising the start box door. In most instances, the

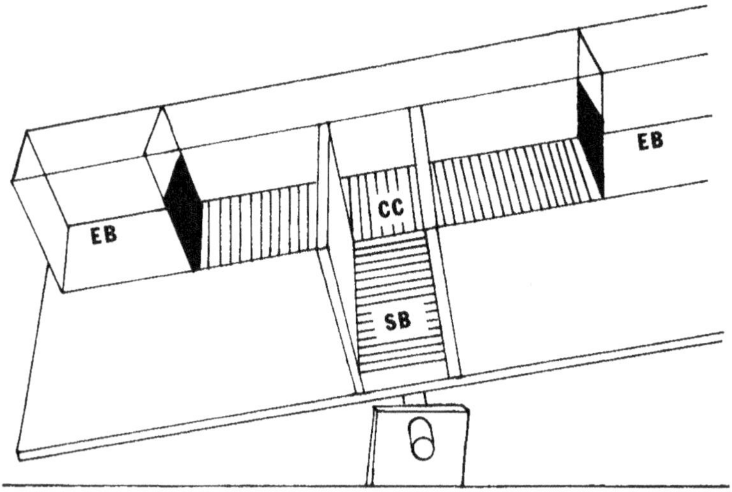

FIGURE 2.5. Schematic drawing of the inclined plane discrimination apparatus showing the start box (SB), choice chamber (CC), and end boxes (EB).

animal would readily leave the start box, traverse the choice chamber, enter the goal box through the centrally located window, and ingest the water or mash. After 10 sec, the animal was carried to a restraining cage to await the next trial. On the 10th trial, the animals were allowed to ingest the water and mash for approximately 200 sec. (This period, however, was shortened by at least 50% in animals that were consistently reluctant to leave the start box, slow in traversing the choice chamber, or hesitant in ingesting the water and mash during the course of preliminary training.)

Problem A was presented on Day 3, Problem B on Day 4, and Problem C on Day 5. Five trials were given on each problem with an intertrial interval of approximately 90–300 sec. The training procedure was the same as that described in preliminary training, except for the addition of recording response latencies–time between raising the start box door and entering (all four paws placed beyond the threshold) the choice chamber. This measure provided an index of the strength of appetitive motivation.

With respect to Problem A, an error consisted of passing under the raised platform by at least the length of the animal's body (excluding the tail). For Problem B, an error consisted of traversing beyond the opening of the elevated cylinder by at least the length of the animal's body, and an error on Problem C consisted of traversing beyond the ladder by at least the length of the animal's body. Total (initial combined with intratrial repetitive) errors were recorded on each trial.

Puzzle-Box Problems

Following a 3-week recovery period, the animals were deprived of water for 2 days and subsequently allowed to explore the entire apparatus for 10 min (Day 1—Monday). During this phase of training, the start box door was in the raised position and the door separating the choice chamber from the goal box was left ajar. A dish of water as well as a dish of sweetened wet mash were available in the goal box from which the animal could freely ingest. On Day 2 (Tuesday), the animals—usually run in squads of two or three—were required to push open the door between the choice chamber and goal box in order to gain access to the water and mash.

Problem 1 was given on Day 3 (Wednesday), Problem 2 on Day 4 (Thursday), and Problem 3 on Day 5 (Friday). At the completion of testing on Day 5, the animals were given continuous access to water in their home cages until Day 6 (Saturday). The remaining five problems were then administered on Days 8–12 (Problem 4 on Monday, Problem 5 on Tuesday, etc.). Ten trials were given on each problem with an intertrial interval of 40–200 sec. A trail commenced when the start box door was raised and usually concluded when the animal successfully opened the door, entered the goal box, and ingested the water/mash for 5 sec. The animal was then removed from the goal box and returned to the holding cage to await the next trial. In the event the animal failed to open the door within 60 sec, the door was partially opened by the experimenter, allowing a space for the rat to enter the goal box.

In an effort to provide some objective measure of motivation, time elapsing from the moment the start box door was raised to the moment the rat made contact with any part of the partition, door, or latch was recorded on every trial.

Problems 2, 3, 5, 6, and 8 were readily learned by the control animals. For these problems, therefore, an error was recorded when the animal failed to open the door within 5 sec after initial contact with any part of the partition, door, or latch. Problems 1, 4, and 7, however, were relatively difficult to solve for the control animals, and, as a consequence, an error was recorded when the animal failed to open the door within 60 sec after making initial contact with any part of the partition, door, or latch.

Aversively Motivated Tasks

The animal was required to learn, in succession, the visual discrimination problem, the maze problem, and finally the inclined plane discrimination problem.

Visual Discrimination

Each rat was allowed to explore the goal box for 5–10 min. The windows were blocked during this time to prevent the animal from entering the choice chamber. Subsequently, each rat was trained to run into the choice chamber from the start box, displace one of the gray cards blocking the window, and enter the goal box in order to escape from (or avoid) foot shock. Training on the brightness discrimination began on the next day. An approach response to the unlocked white card (positive) admitted the animal to the goal box. On the other hand, an approach response to the adjacent locked black card (negative) was punished by foot shock, the animal subsequently being forced to respond to the white card in order to gain entrance into the goal box. An error was defined as an approach response to the black card that brought the animal's forefeet in contact with the charged grid section that extended 8.0 cm in front of the negative card. The position of the positive (and negative) card was switched from the right to the left side in a sequence mixed with single- and double-alternation runs. (In our experience, this sequence reduces the incidence of position habits in brain-damaged and control animals.) Usually, 8–12 trials were given daily, with intertrial interval of 60–90 sec. The criterion of learning consisted of the first appearance of either a "perfect" or "near-perfect" run of correct responses having a probability of occurrence of less than 0.05 (Runnels, Thompson, & Runnels, 1968), followed by at least 75% correct responses in the subsequent block of eight trials given on the next day.

The specific training procedure was as follows: The animal was placed in the start box and the start box door was raised. Failure to leave the start box within 5 sec was followed by interrupted foot shocks until the animal entered the choice chamber. No further foot shocks were given unless the animal made an error or failed to respond to one of the cards within 5 sec. The animal was allowed to remain in the goal box for 10 sec, after which it was carried to the restraining cage to await the next trial. The animals were usually run in squads of two or three.

Maze

Each animal was first habituated to the goal box for 5 min. Subsequently, it was placed in the start box and the start box door was raised. Failure to leave the start box within 5 sec was followed by foot shocks until the animal entered the maze proper. No further foot shocks were given unless the animal made an error (entered a blind alley by at least the length of its head and thorax) or stopped forward progression in the maze. Upon entering the goal box, a guillotine door was lowered to prevent reentry into the maze. The animal was permitted to remain in the goal box for 10 sec, after which it was transferred to the restraining cage to await the next trial. Four trials were usually given each day, with an intertrial interval of 60–90 sec. Training was terminated when the animal made no more than one error in eight consecutive trials. The animals were usually run in squads of two or three.

Inclined Plane Discrimination

The animals were first habituated to each end box for 5 min and then were trained to run into one of the arms from the start box, displace the card blocking the window, and enter the end box in order to escape from (or avoid) foot shock. During this period, the apparatus was in the horizontal position. One hour later, the animals were blinded by enucleation (under deep chloral hydrate anesthesia) in order to ensure that subsequent learning of this discrimination was based on nonvisual cues.

On the following day, the animals were trained on the vestibular–proprioceptive–kinesthetic discrimination. An approach response to the upward sloping arm led to an end box that could be entered by displacing the unlocked card, whereas an approach response to the downward sloping arm led to a locked card preventing the animal from entering the end box on that side. Punishment for an error (approach to within 10.8 cm of the locked card) was given by charging the grid section located below the locked card. The position of the correct (and incorrect) arm varied from the right side to the left in a sequence mixed with single- and double-alternation runs and 8–12 trials were given daily, with an intertrial interval of 60–90 sec. The criterion of learning and the specific training procedure were the same as those described for the visual discrimination task.

Histology

At the conclusion of postoperative training, each brain-damaged animal was killed with an overdose of chloral hydrate, its vascular system usually perfused with normal saline followed by 10% formalin, and the brain removed and stored in 10% formalin for 2–4 days. The cortical lesions were reconstructed on Lashley-type brain diagrams, from which the percentage of neocortical destruction was determined by a procedure similar to that used by Lashley (1929). Each brain with cortical or subcortical lesions was blocked, frozen, and sectioned frontally at 90 μm. Every 3rd or 4th section through the lesion was retained and subsequently photographed at 12× by using the section as a negative film in an enlarger. These photographs of unstained sections yield differentiation of the brain field similar to that obtained with a fiber stain and readily permit identification of the three zones of the electrolytic lesion—the vacuolated area, the narrow rim of coagulated tissue, and the surrounding gliosis (see Figure 2.6). Throughout this monograph, however, only the first two zones were considered in describing the locus and extent of the lesions, since the area of gliosis may contain some normal cells and intact fiber tracts.

Constructing the Atlas

The results of this lesion survey have been depicted on two separate sets of maps. One set deals with the pattern of learning deficits associated with each lesion placement ("Learning Syndromes"), and the second focuses on the distribution of lesion placements associated with a learning deficit on a specif-

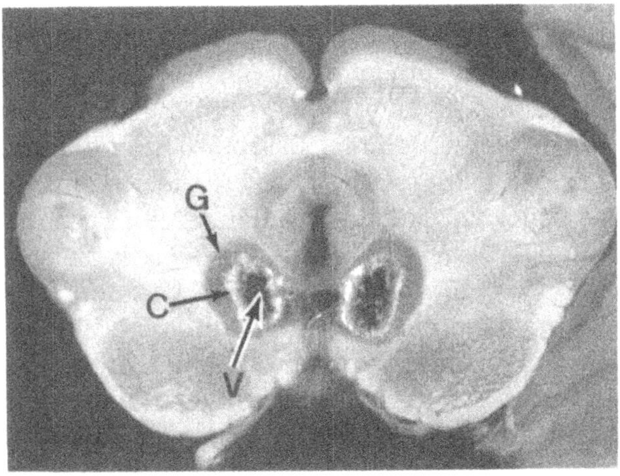

FIGURE 2.6. Unstained brain section showing an electrolytic lesion in the region of the red nucleus. Note the three zones of the lesion—vacuolated area (V), narrow rim of severely coagulated tissue (C), and gliosis (G). The central necrotic zone of the lesion consists of the areas occupied by V and C only.

ic task ("Learning Systems"). The criterion for determining the presence of a specific learning deficit associated with a particular lesion placement was based on the difference in mean learning scores between the brain-damaged group and its corresponding sham-operated control group. If the mean learning (error) score of the former was found to be significantly larger than that of the latter ($p < 0.05$, Mann–Whitney U test, two-tailed), then a learning deficit was considered to exist and was included within the syndrome characteristic of the lesion sustained by the brain-damaged group in question.

Maps of Learning Syndromes

These maps show the site of a "typical" lesion placement to a given structure and its associated syndrome on a line drawing of a brain section. In order to provide an indication as to the magnitude of the deficit observed in the sample of brain-damaged animals examined, the following procedure was used: A moderate deficit (mean error score of the brain-damaged group was less than twice as large as that of the corresponding control group) was listed under the syndrome in plain lowercase letters, while a severe deficit (mean error score was at least twice as large as that of the corresponding control group) was denoted in underscored lowercase letters.

Accompanying each line drawing is a photograph of a corresponding unstained brain section appropriately labeled and, in the case of the frontal sections, furnished with frontal, lateral, and vertical stereotaxic coordinates based on a modified version of the Massopust rat atlas (see Thompson, 1978a). These illustrations are organized in a manner similar to that of a stereotaxic atlas. The first two diagrams present a dorsal and lateral view of the cerebral surface along with the site of the lesions to the cerebral cortex, olfactory bulbs, and cerebellum. (The eye orbits are shown to indicate that one of the lesions involved enucleation of the eyes.) These are followed by a diagram of a parasagittal section of the brain showing the locus and extent of damage to two separate regions of the mesial (limbic) cortex. In subsequent illustrations, diagrams of frontal

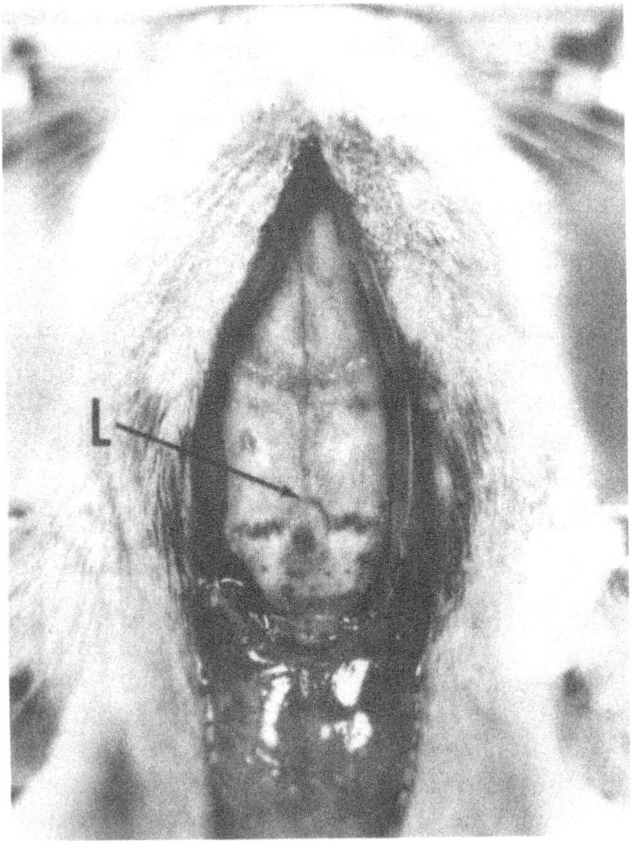

FIGURE 2.7. Exposed skull and neck muscles of the rat, showing the position of lambda (L).

sections through the brainstem are used to depict the sites of lesions to the various structures located within the interior parts of the brain. These frontal sections proceed from anterior to posterior.

The modified version of the Massopust rat atlas is presented in the form of enlarged photographs of frontal sections of the brain cut in the vertical plane. The lateral and vertical coordinates are scaled in millimeters. At the top of each photograph, F refers to the level of the frontal section in terms of the distance in millimeters from lambda (see Figure 2.7). Coordinate measurements caudal to lambda are indicated with a minus (−) sign; those rostral to lambda are given in positive numbers. To avoid confusion and prevent visual overcrowding in the photographs, identification of brain structures was achieved by lettering (abbreviations) and was limited to those 50 sites selected as targets for our lesions. The specific naming of the nuclei did not follow any particular convention. In most cases, however, the nomenclature was guided by the rat atlases of Massopust (1961) and Paxinos and Watson (1982).

It is important to emphasize that the values of the coordinates shown on these maps are based on albino rats weighing 50–65 g. These values were used to guide the lesion electrode to those subcortical targets intended for destruction.

Maps of Learning Systems

These maps illustrate the distribution of those lesion placements producing learning deficits on a particular problem-solving task. One map relates to the visual discrimination problem, the second to the maze problem, the third to the inclined plane discrimination problem, and the fourth to Detour Problem A. (Maps of Detour Problems B and C are not included because they would be confounded by transfer effects resulting from prior acquisition of Detour Problem A.) These maps consist of a parasagittal section of the rat brain, showing the relative positions of lesion placements producing moderate (light shaded areas) and severe (heavy shaded areas) acquisition deficits.

3

Lesion Placements and Learning Syndromes

List of Lesions and Corresponding Illustrations

Area damaged	Illustration number
E = Eyes	3.2
Neocortex	
FC = Frontal cortex (dorsal region)	3.1
VF = Frontal cortex (ventral region)	3.2
PA = Parietal cortex	3.1
OC = Occipitotemporal cortex	3.1
Other telencephalic areas	
OL = Olfactory bulbs	3.2
AC = Frontocingulate cortex	3.3
PC = Cingulate cortex (posterior region)	3.3
ES = Entorhinosubicular area	3.13
NA = Nucleus accumbens septi	3.4
SF = Septofornix area	3.5
DH = Hippocampus (dorsal)	3.9

Area damaged	Illustration number
VH = Hippocampus (ventral)	3.11
AM = Amygdala	3.8
RC = Caudatoputamen (rostral)	3.4
DC = Caudatoputamen (dorsal)	3.5
VC = Caudatoputamen (ventral)	3.5
GP = Globus pallidus	3.6
VP = Ventral pallidum	3.6
EP = Entopeduncular nucleus	3.8
Thalamus	
AT = Anterior complex	3.7
VL = Ventrolateral complex	3.8
CL = Centrolateral region	3.8
LT = Lateral complex	3.9
MD = Mediodorsal complex	3.9
VM = Ventromedial region	3.9
VB = Ventrobasal region	3.10
P = Parafascicular region	3.10
H = Habenular nuclei	3.9
Hypothalamus	
HA = Anterior region	3.7
HV = Ventromedial region	3.9
LH = Posterolateral region	3.10
MB = Mammillary bodies	3.10
ST = Subthalamus	3.10
Pretectal area	
PT = Medial pretectal area	3.10
AP = Anterior pretectal nucleus (nucleus posterior)	3.10

Area damaged	Illustration number
Brainstem reticular formation	
MF = Midbrain area (paramedial)	3.11
PF = Pontine area (paramedial)	3.15
Other brainstem structures	
SC = Superior colliculus	3.12
IC = Inferior colliculus	3.15
CG = Central gray (midbrain)	3.12
VT = Ventral tegmental area	3.11
SN = Substantia nigra (lateral)	3.11
RN = Red nucleus area	3.12
IP = Interpedunculocentral tegmental area	3.12
MR = Raphe area (median)	3.14
DR = Raphe area (dorsal)	3.13
LM = Lateral midbrain area	3.14
PN = Pedunculopontine area (dorsal)	3.14
CB = Cerebellum	3.2

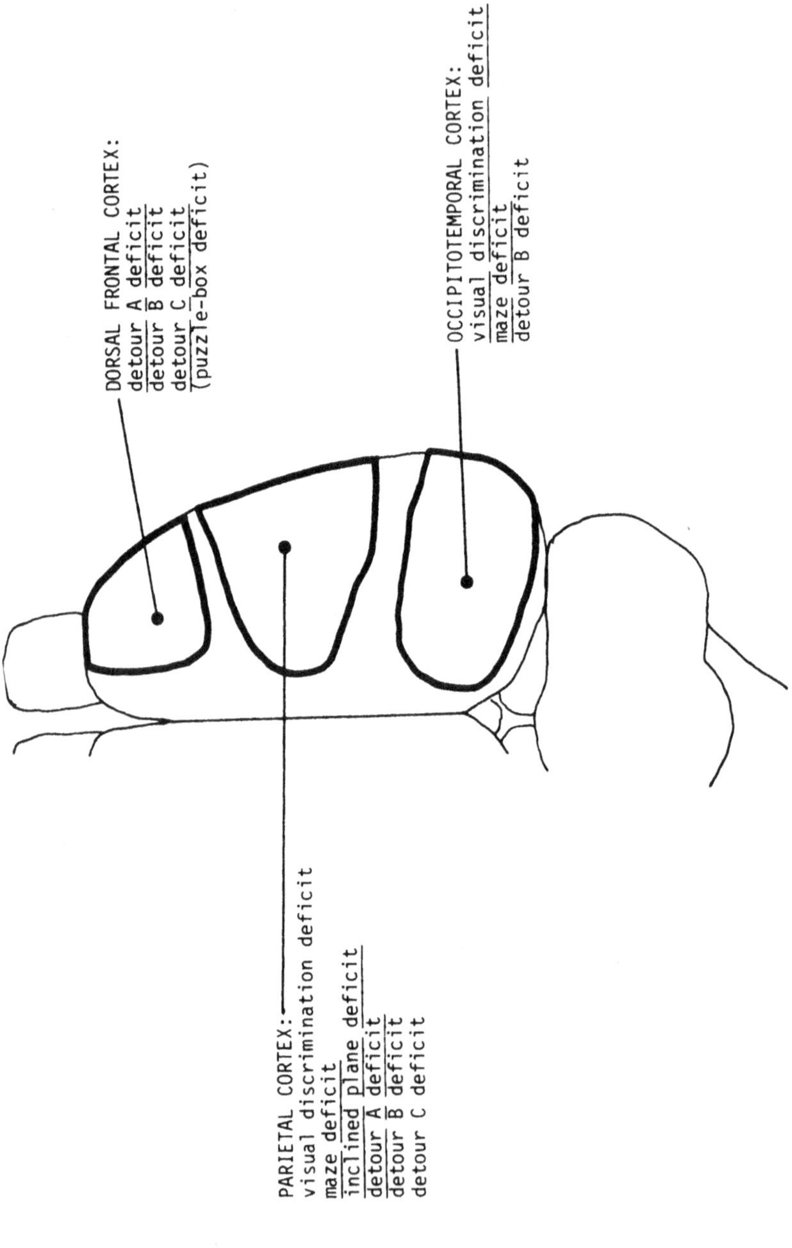
DORSAL FRONTAL CORTEX:
detour A deficit
detour B deficit
detour C deficit
(puzzle-box deficit)
OCCIPITOTEMPORAL CORTEX:
visual discrimination deficit
maze deficit
detour B deficit
PARIETAL CORTEX:
visual discrimination deficit
maze deficit
inclined plane deficit
detour A deficit
detour B deficit
detour C deficit

FIGURE 3.1A

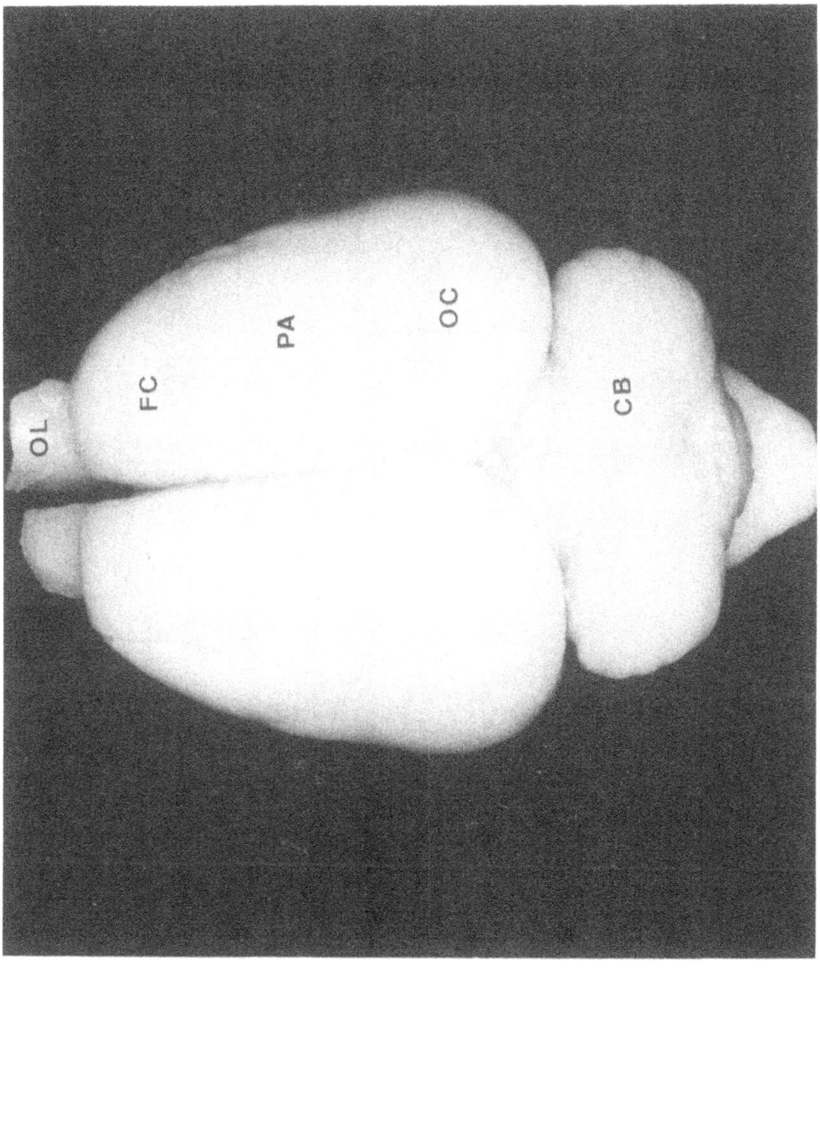

FIGURE 3.1B

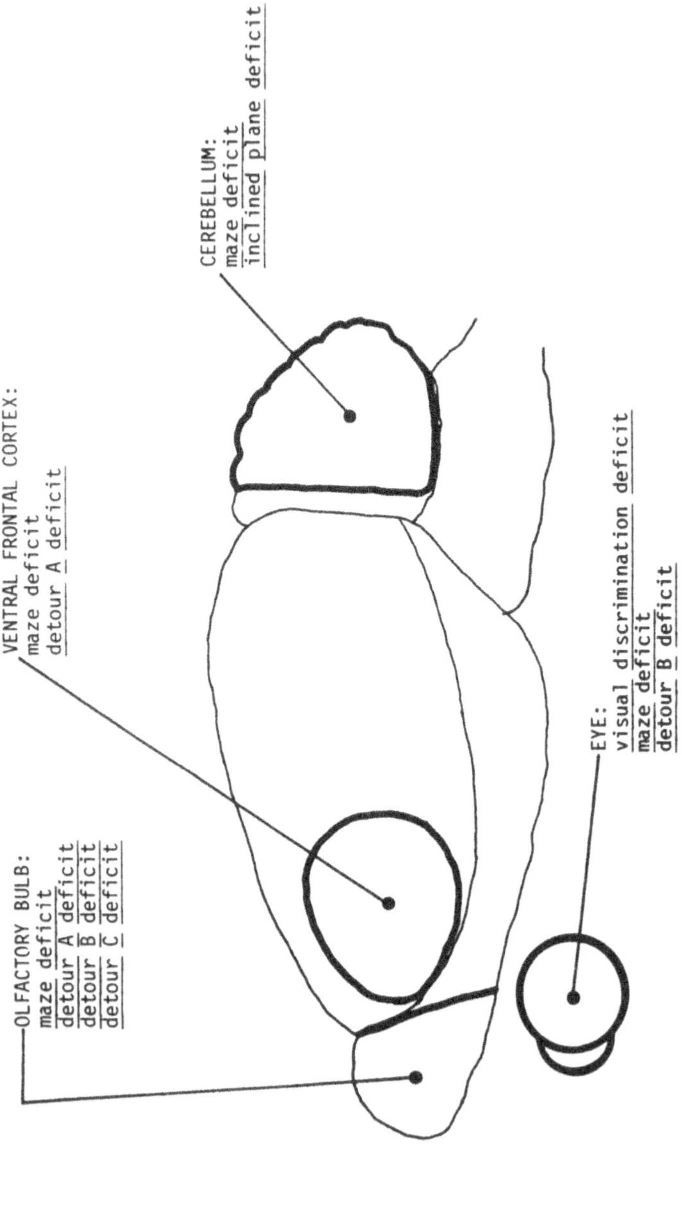
CEREBELLUM:
maze deficit
inclined plane deficit
VENTRAL FRONTAL CORTEX:
maze deficit
detour A deficit
EYE:
visual discrimination deficit
maze deficit
detour B deficit
OLFACTORY BULB:
maze deficit
detour A deficit
detour B deficit
detour C deficit

FIGURE 3.2A

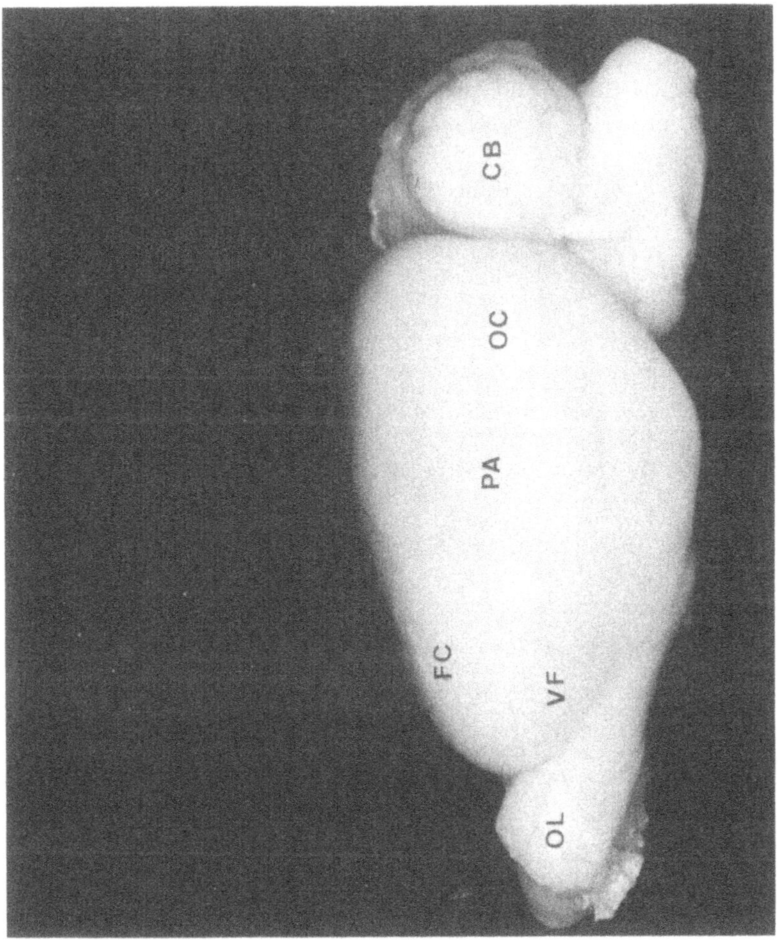

FIGURE 3.2B

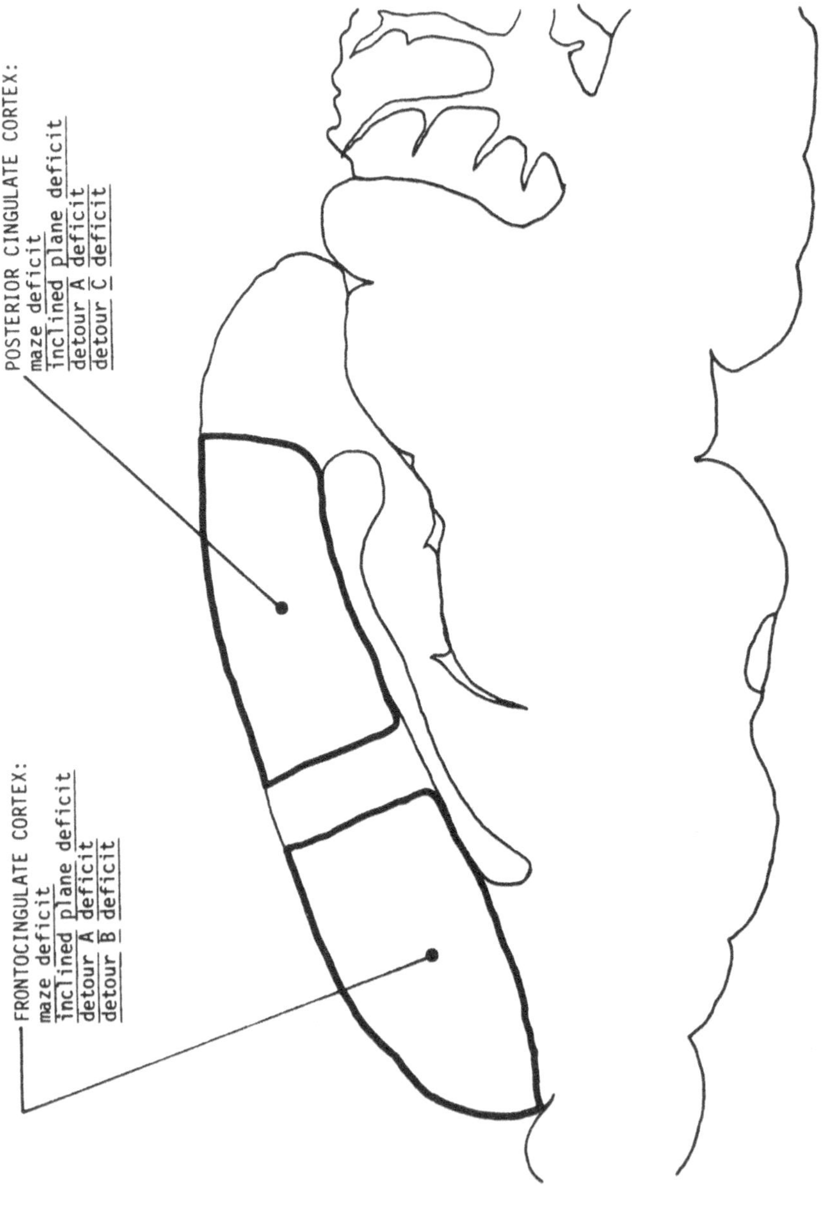
POSTERIOR CINGULATE CORTEX:
maze deficit
inclined plane deficit
detour A deficit
detour C deficit
FRONTOCINGULATE CORTEX:
maze deficit
inclined plane deficit
detour A deficit
detour B deficit

FIGURE 3.3A

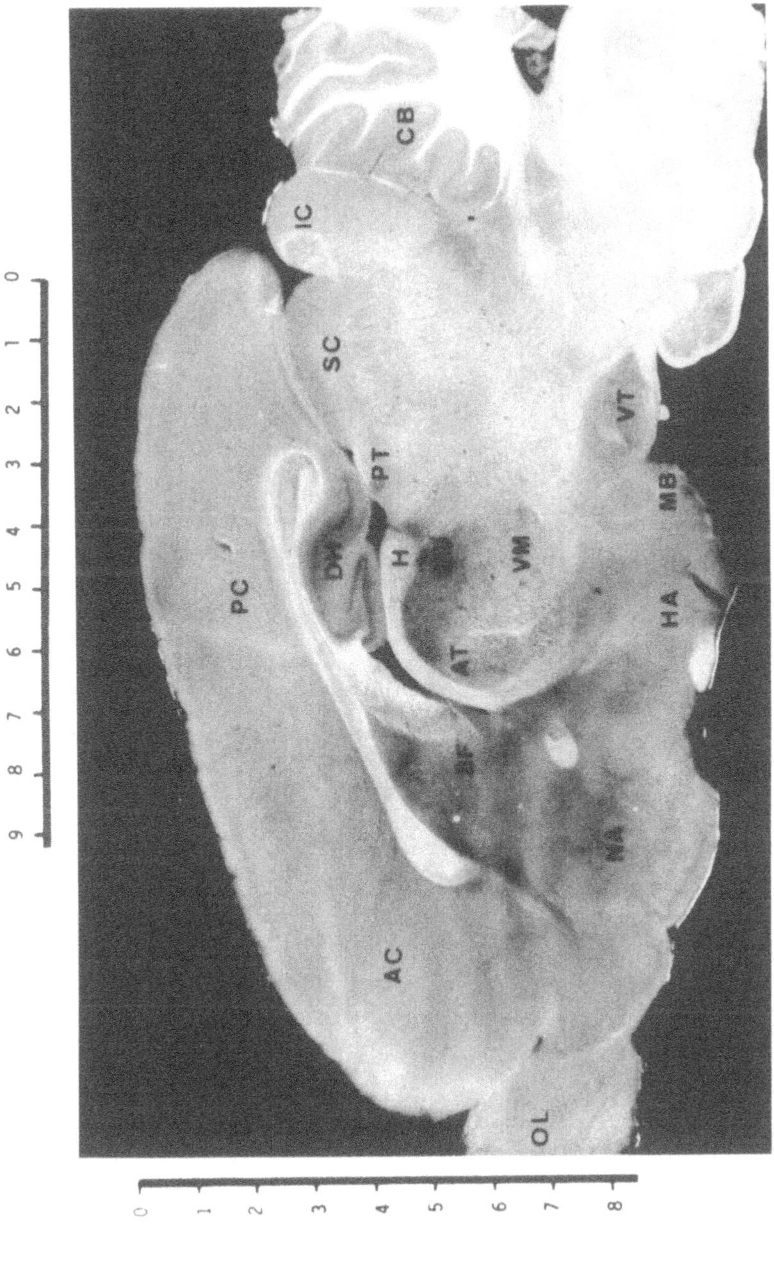

FIGURE 3.3B

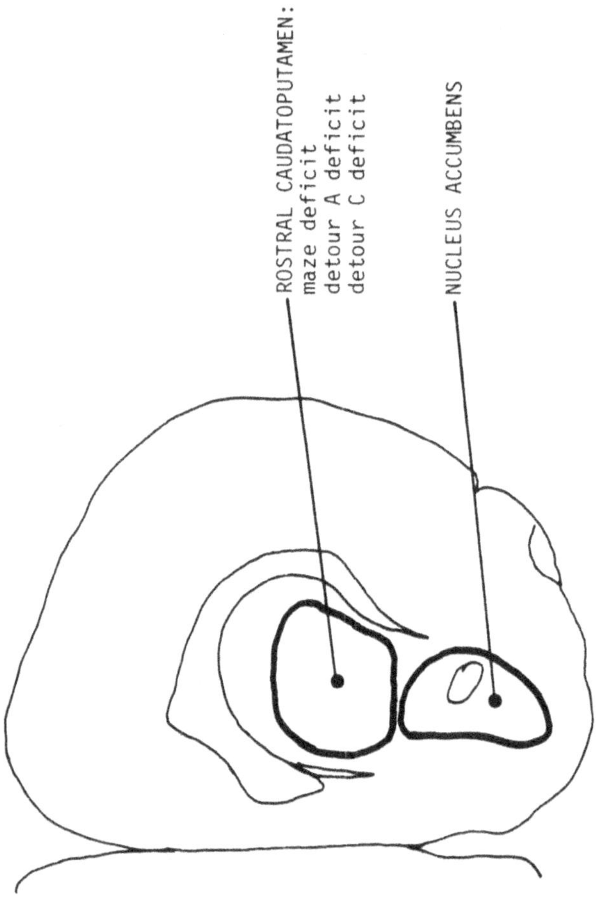
ROSTRAL CAUDATOPUTAMEN:
maze deficit
detour A deficit
detour C deficit
NUCLEUS ACCUMBENS

FIGURE 3.4A

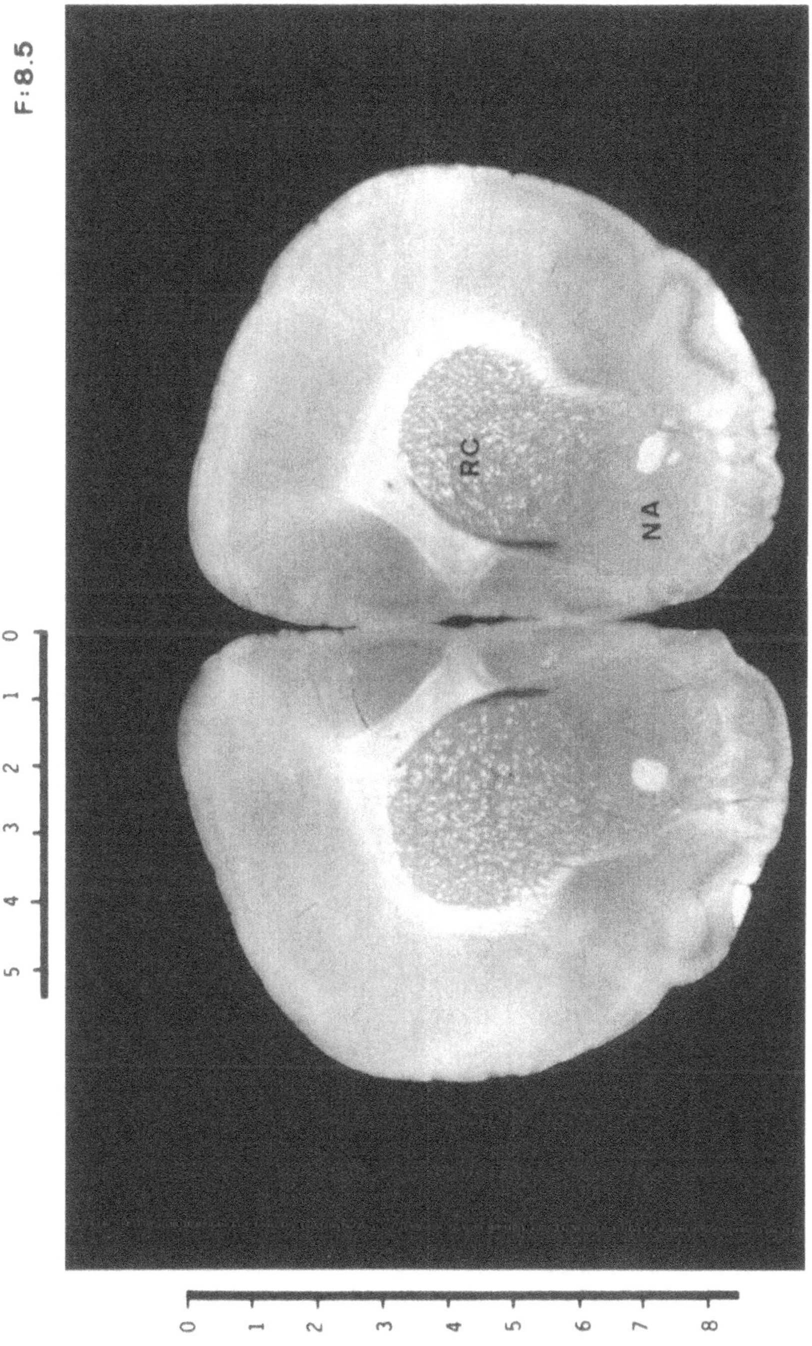

FIGURE 3.4B

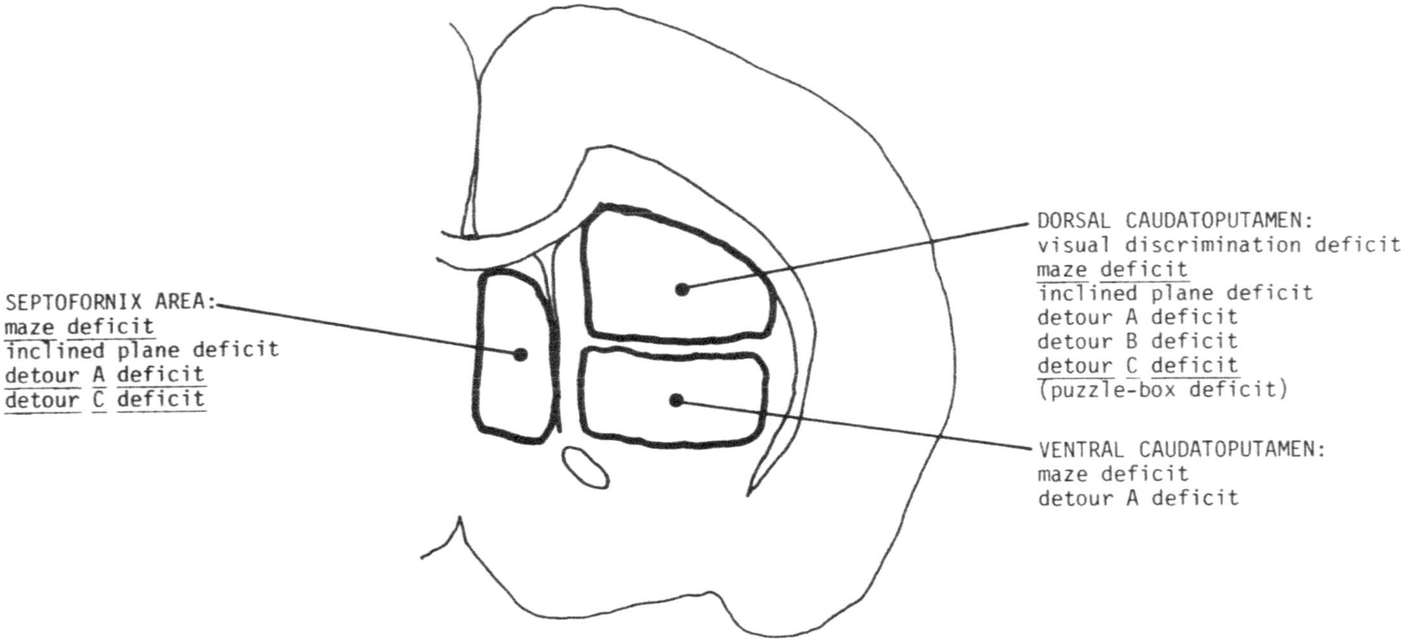
SEPTOFORNIX AREA:
maze deficit
inclined plane deficit
detour A deficit
detour C deficit
DORSAL CAUDATOPUTAMEN:
visual discrimination deficit
maze deficit
inclined plane deficit
detour A deficit
detour B deficit
detour C deficit
(puzzle-box deficit)
VENTRAL CAUDATOPUTAMEN:
maze deficit
detour A deficit

FIGURE 3.5A

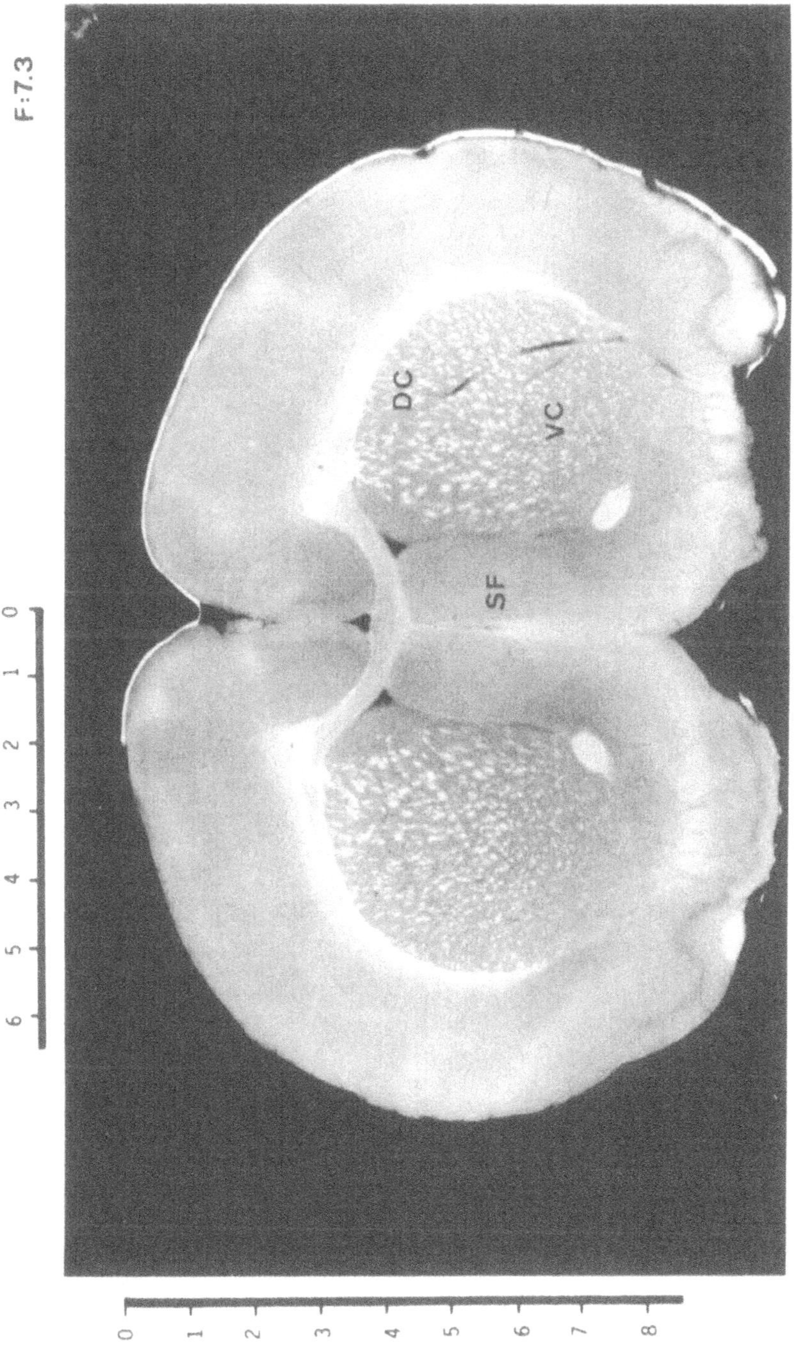

FIGURE 3.5B

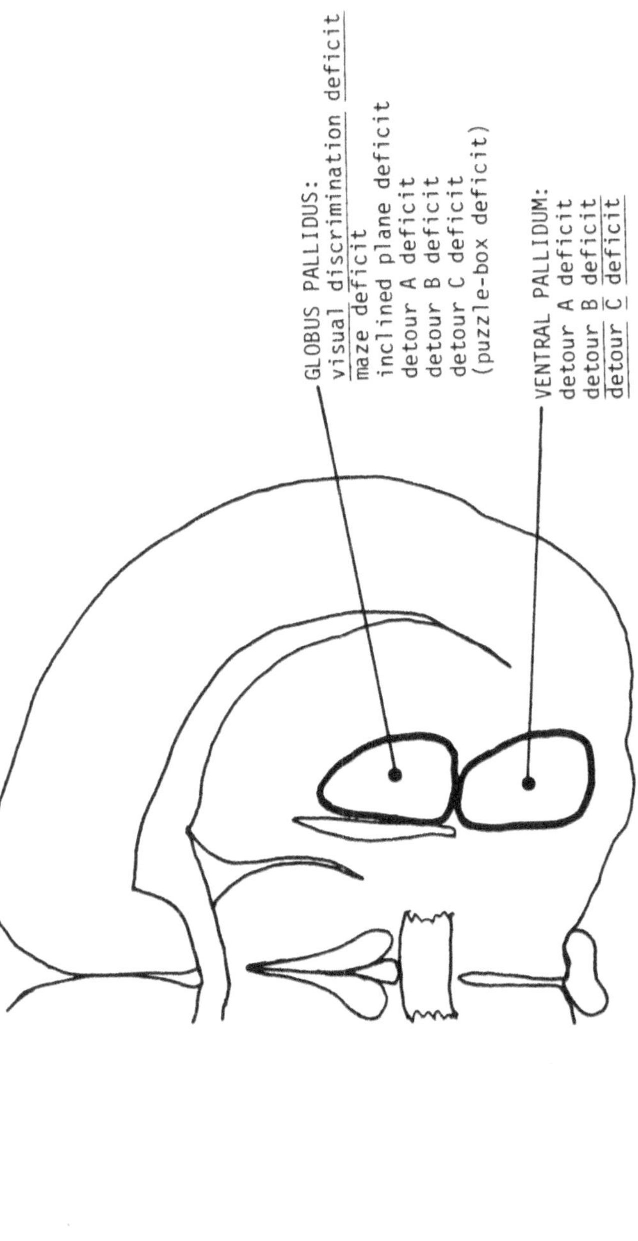
GLOBUS PALLIDUS:
visual discrimination deficit
maze deficit
inclined plane deficit
detour A deficit
detour B deficit
detour C deficit
(puzzle-box deficit)
VENTRAL PALLIDUM:
detour A deficit
detour B deficit
detour C deficit

FIGURE 3.6A

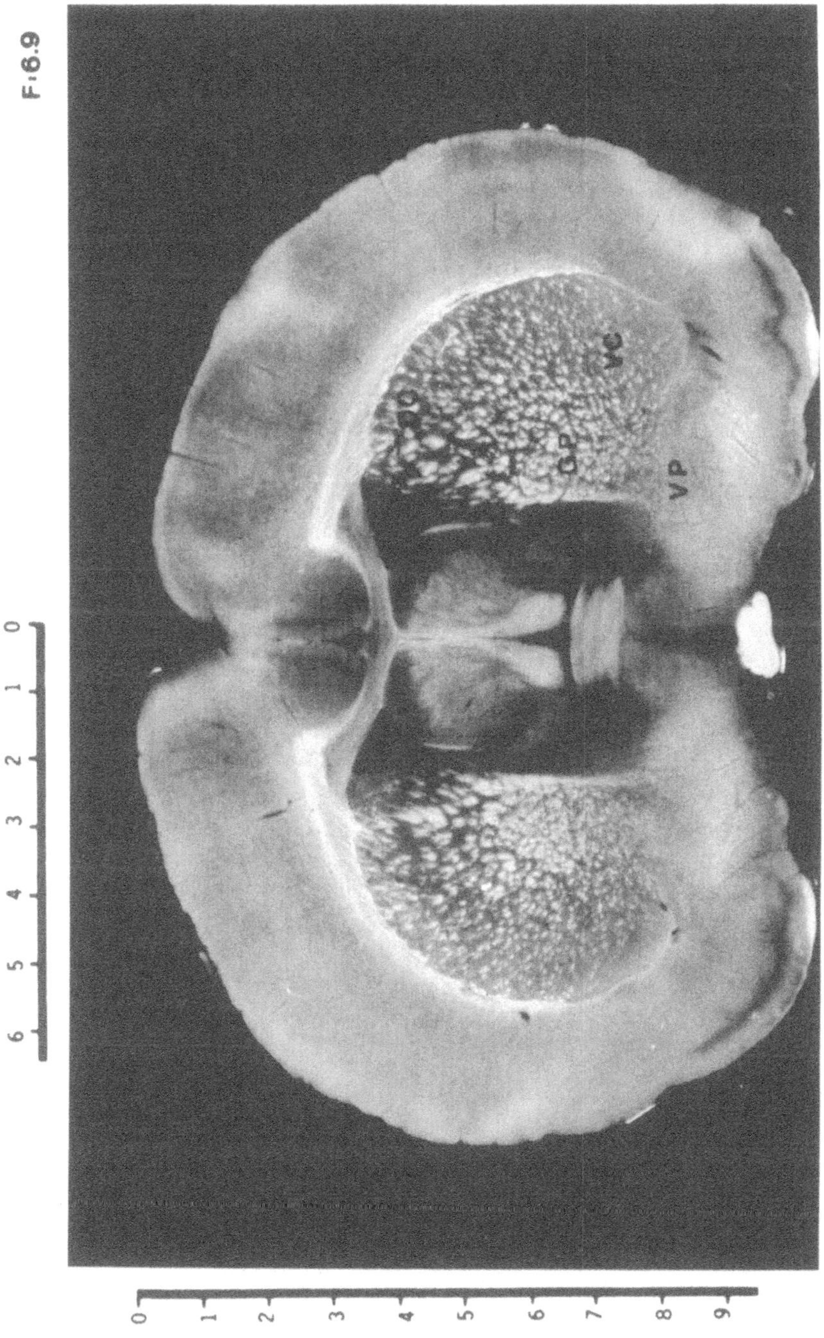

FIGURE 3.6B

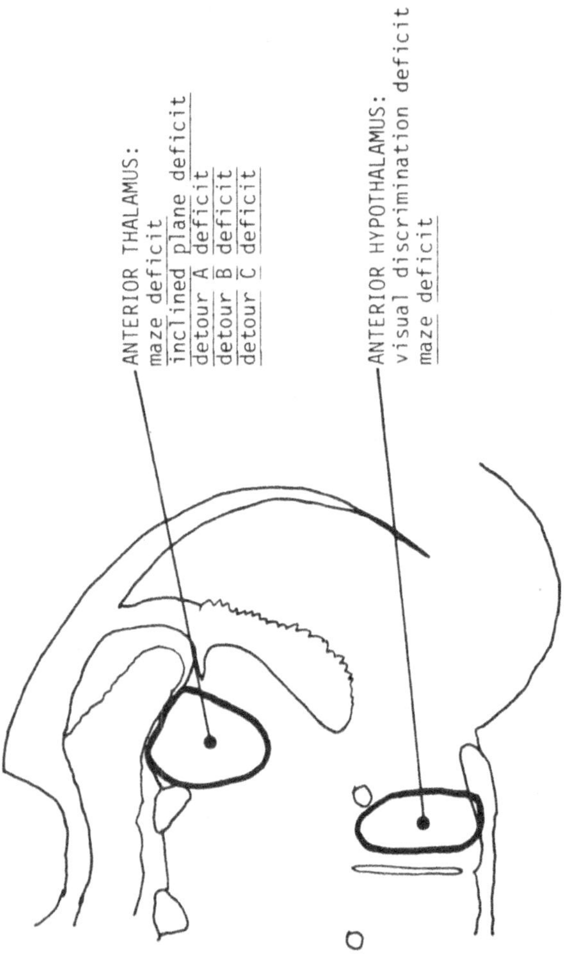
ANTERIOR THALAMUS:
maze deficit
inclined plane deficit
detour A deficit
detour B deficit
detour C deficit
ANTERIOR HYPOTHALAMUS:
visual discrimination deficit
maze deficit

FIGURE 3.7A

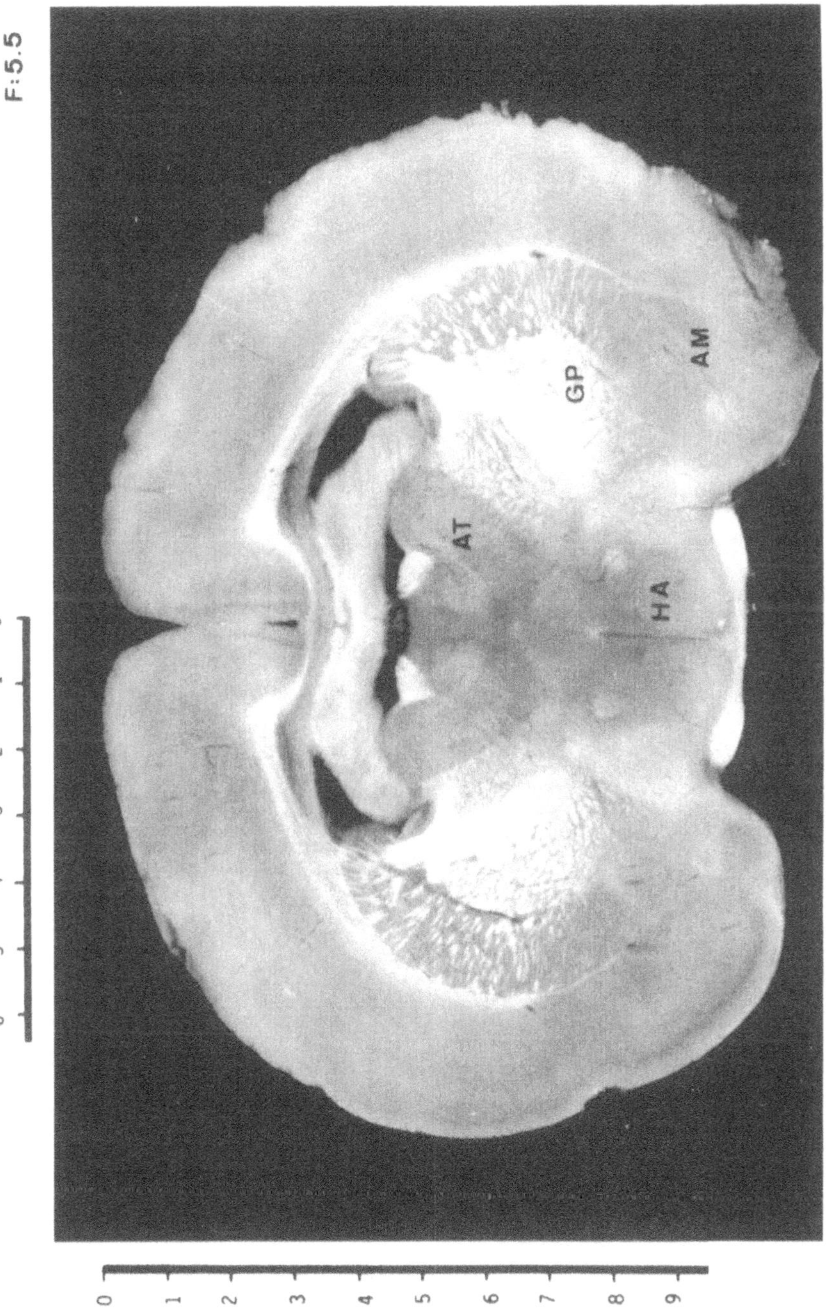

FIGURE 3.7B

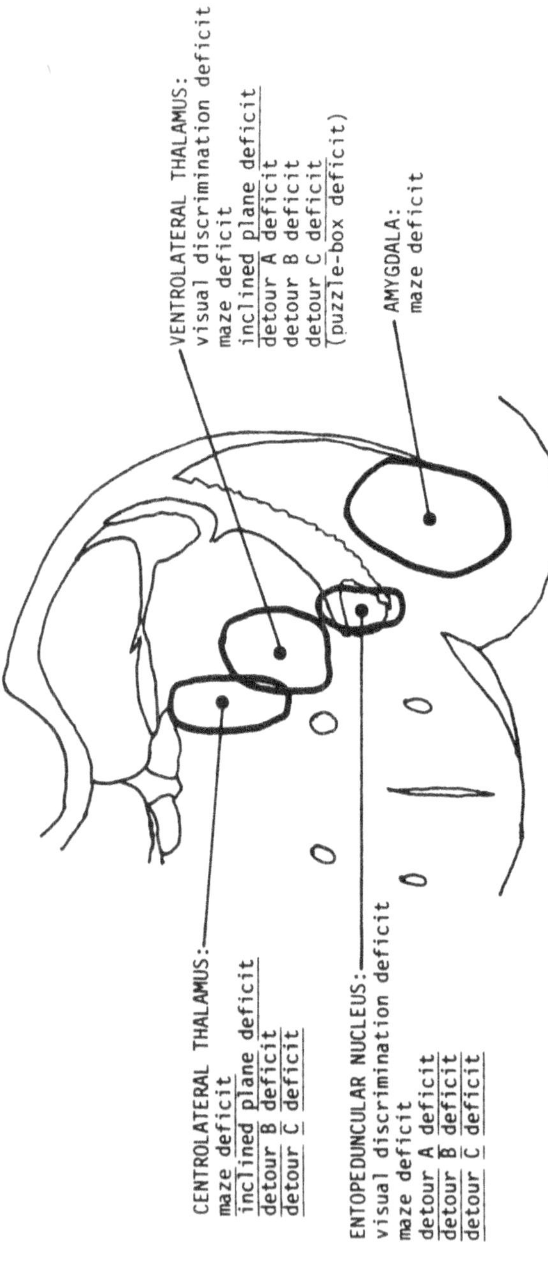
VENTROLATERAL THALAMUS:
visual discrimination deficit
maze deficit
inclined plane deficit
detour A deficit
detour B deficit
detour C deficit
(puzzle-box deficit)
AMYGDALA:
maze deficit
CENTROLATERAL THALAMUS:
maze deficit
inclined plane deficit
detour B deficit
detour C deficit
ENTOPEDUNCULAR NUCLEUS:
visual discrimination deficit
maze deficit
detour A deficit
detour B deficit
detour C deficit

FIGURE 3.8A

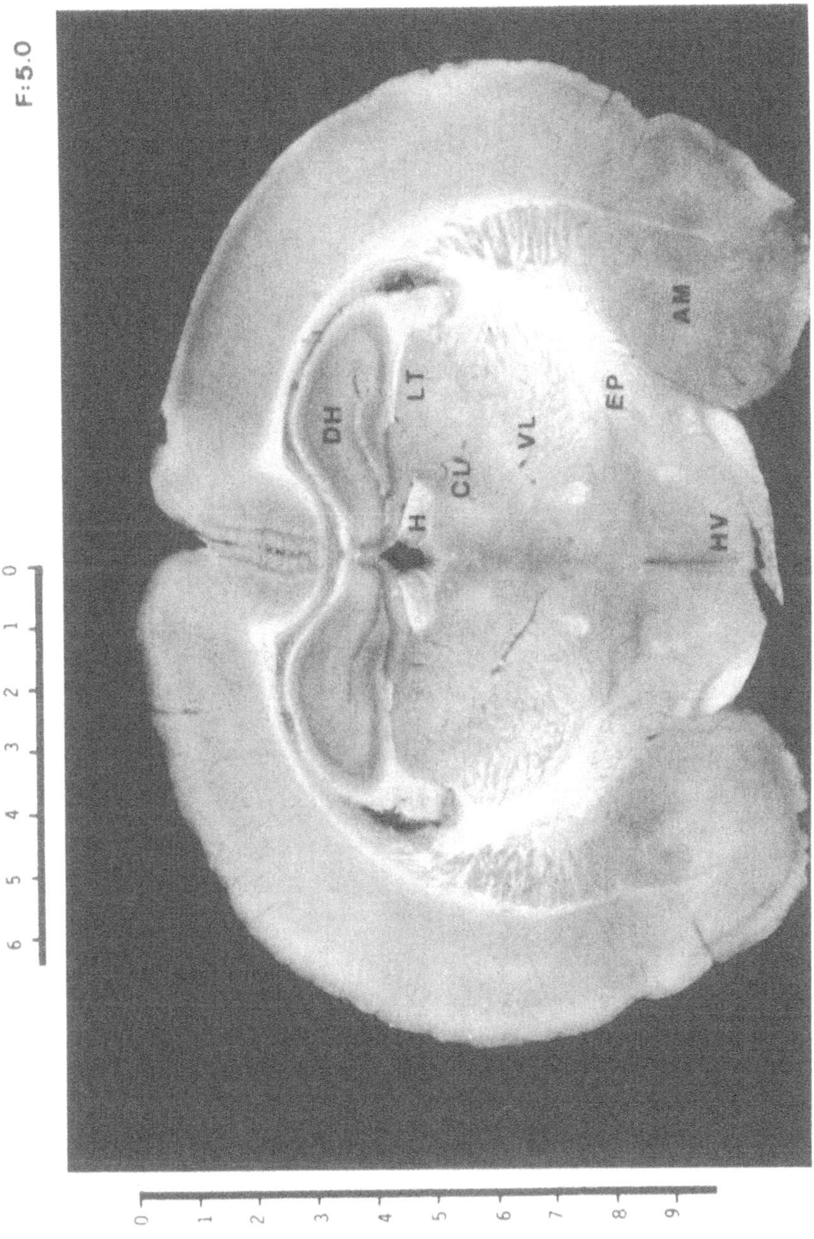

FIGURE 3.8B

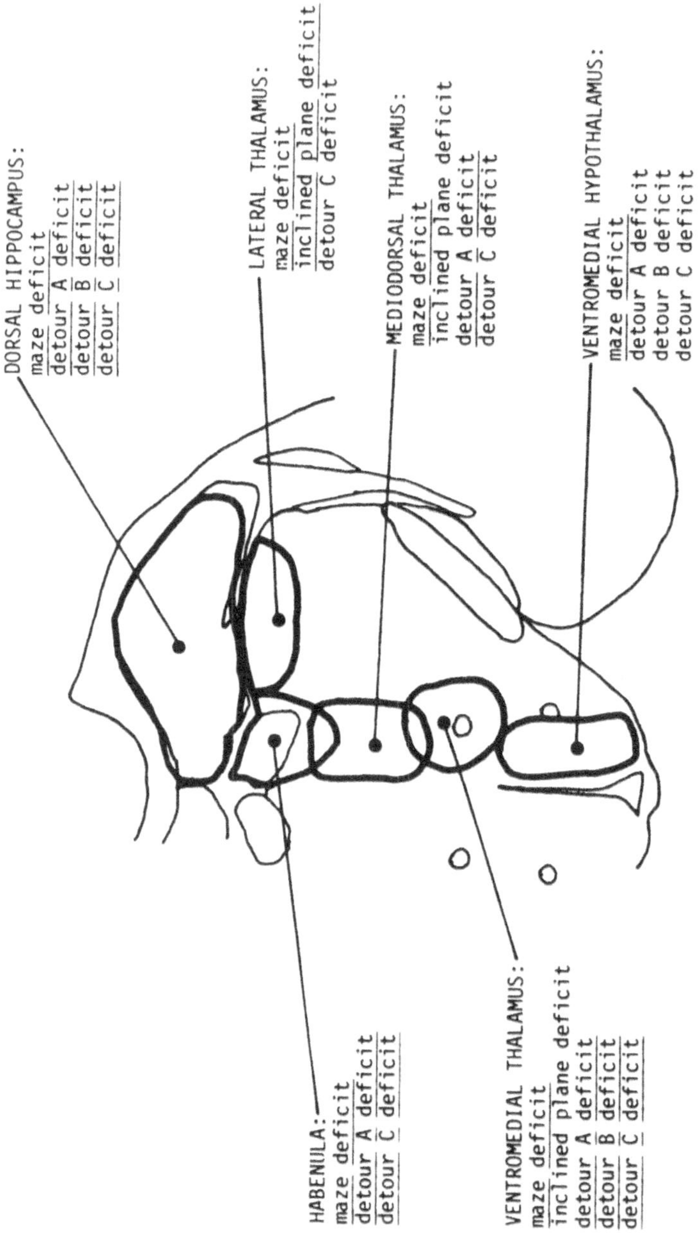
DORSAL HIPPOCAMPUS:
maze deficit
detour A deficit
detour B deficit
detour C deficit
LATERAL THALAMUS:
maze deficit
inclined plane deficit
detour C deficit
MEDIODORSAL THALAMUS:
maze deficit
inclined plane deficit
detour A deficit
detour C deficit
VENTROMEDIAL HYPOTHALAMUS:
maze deficit
detour A deficit
detour B deficit
detour C deficit
HABENULA:
maze deficit
detour A deficit
detour C deficit
VENTROMEDIAL THALAMUS:
maze deficit
inclined plane deficit
detour A deficit
detour B deficit
detour C deficit

FIGURE 3.9A

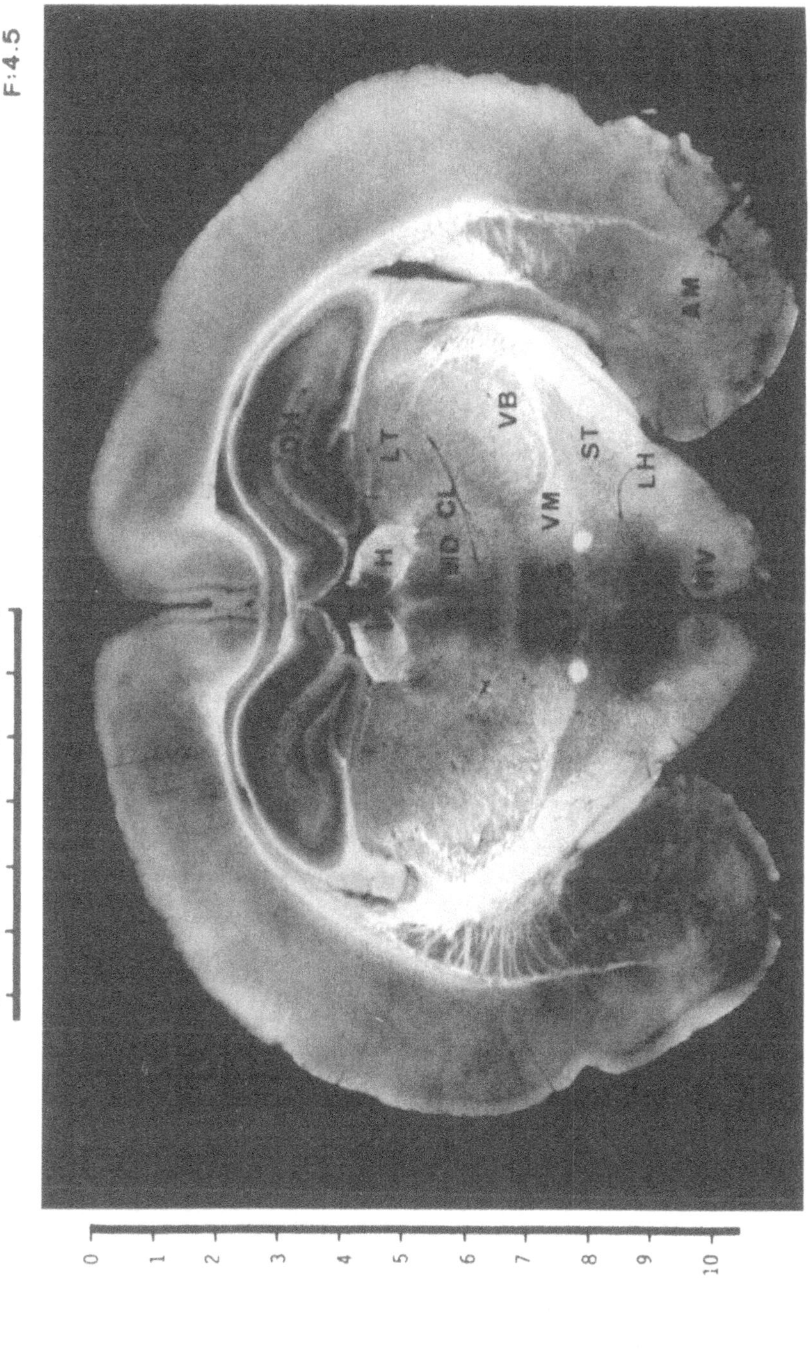

FIGURE 3.9B

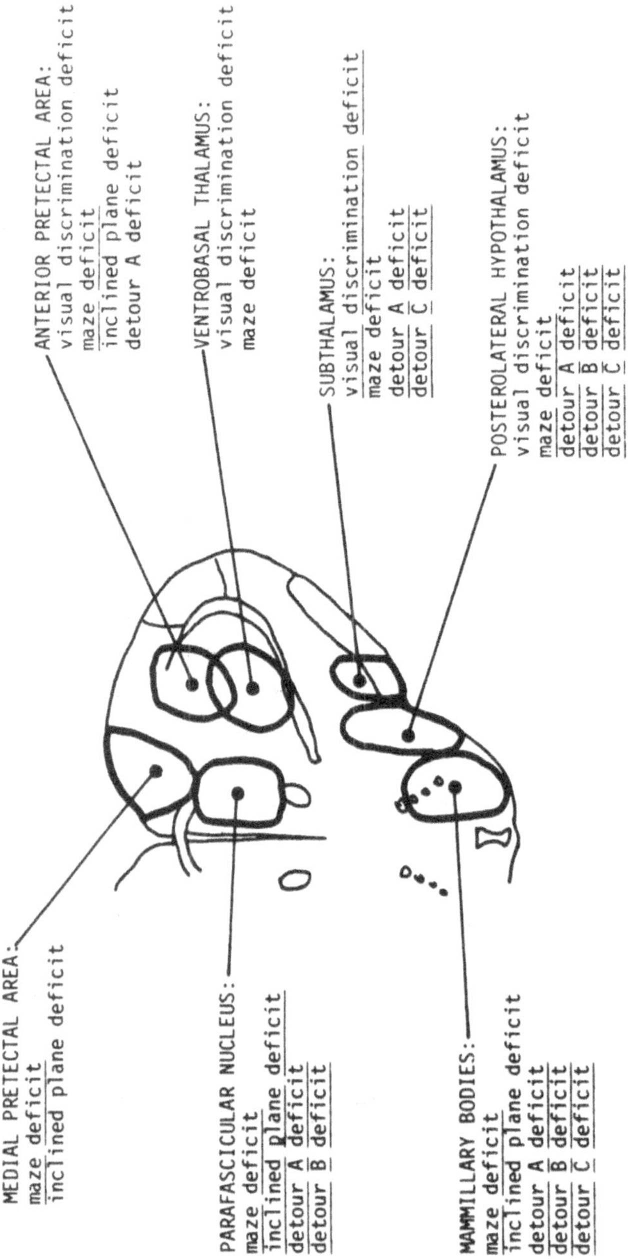

MEDIAL PRETECTAL AREA:
maze deficit
inclined plane deficit
ANTERIOR PRETECTAL AREA:
visual discrimination deficit
maze deficit
inclined plane deficit
detour A deficit
PARAFASCICULAR NUCLEUS:
maze deficit
inclined plane deficit
detour A deficit
detour B deficit
VENTROBASAL THALAMUS:
visual discrimination deficit
maze deficit
SUBTHALAMUS:
visual discrimination deficit
maze deficit
detour A deficit
detour C deficit
MAMMILLARY BODIES:
maze deficit
inclined plane deficit
detour A deficit
detour B deficit
detour C deficit
POSTEROLATERAL HYPOTHALAMUS:
visual discrimination deficit
maze deficit
detour A deficit
detour B deficit
detour C deficit

FIGURE 3.10A

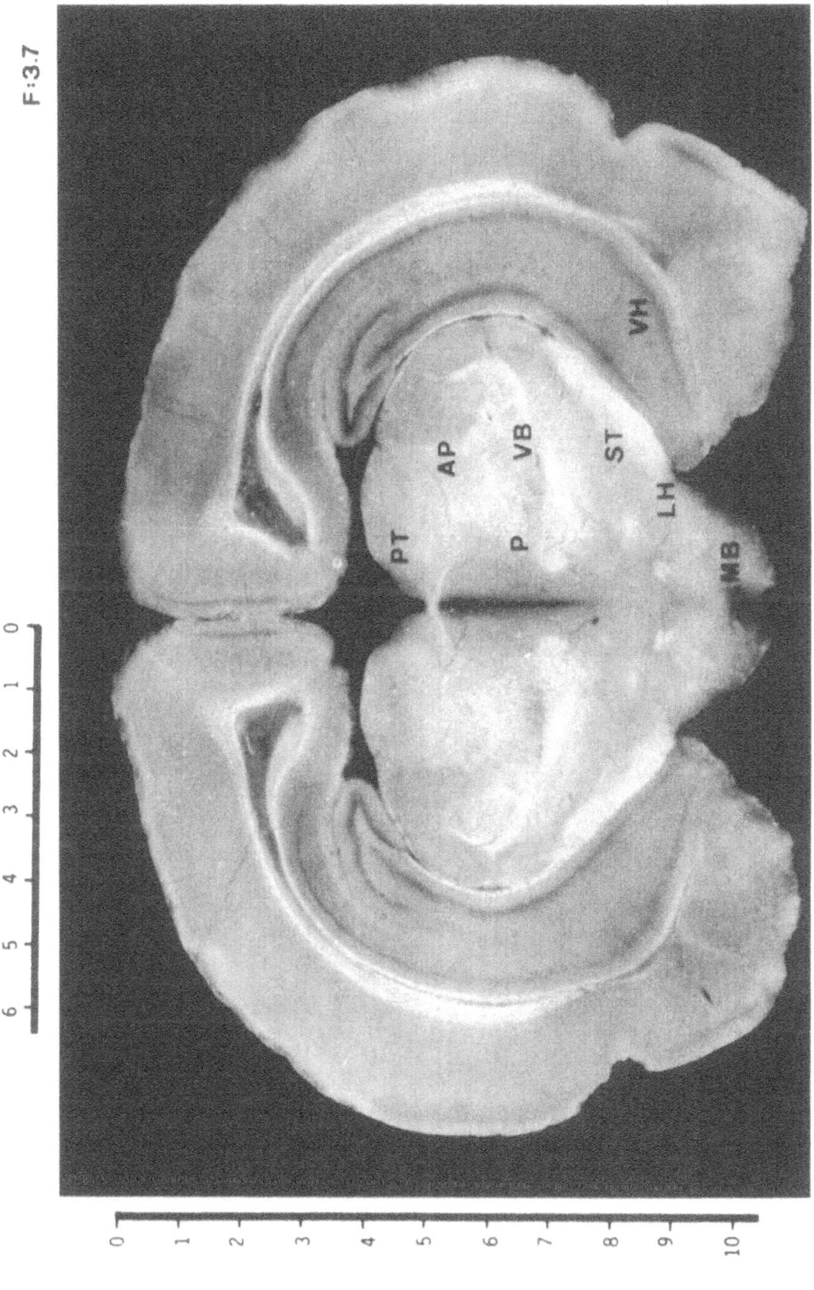

FIGURE 3.10B

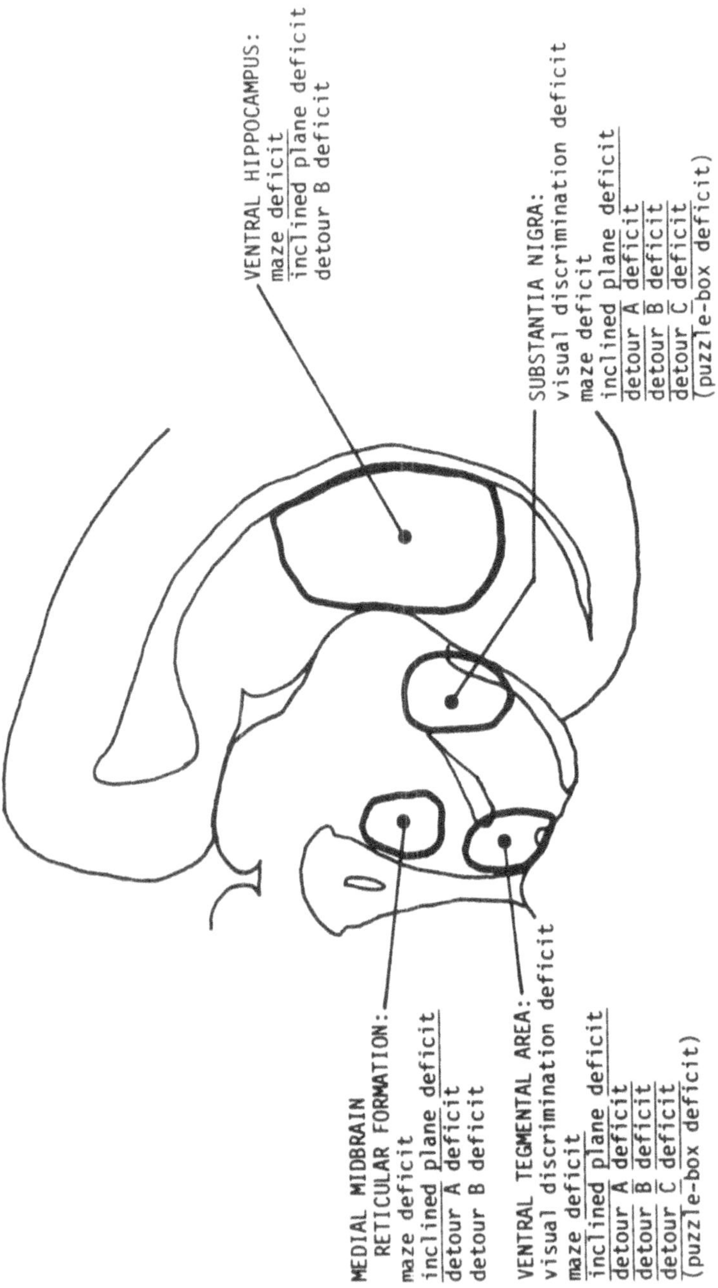
VENTRAL HIPPOCAMPUS:
maze deficit
inclined plane deficit
detour B deficit
SUBSTANTIA NIGRA:
visual discrimination deficit
maze deficit
inclined plane deficit
detour A deficit
detour B deficit
detour C deficit
(puzzle-box deficit)
MEDIAL MIDBRAIN
RETICULAR FORMATION:
maze deficit
inclined plane deficit
detour A deficit
detour B deficit
VENTRAL TEGMENTAL AREA:
visual discrimination deficit
maze deficit
inclined plane deficit
detour A deficit
detour B deficit
detour C deficit
(puzzle-box deficit)

FIGURE 3.11A

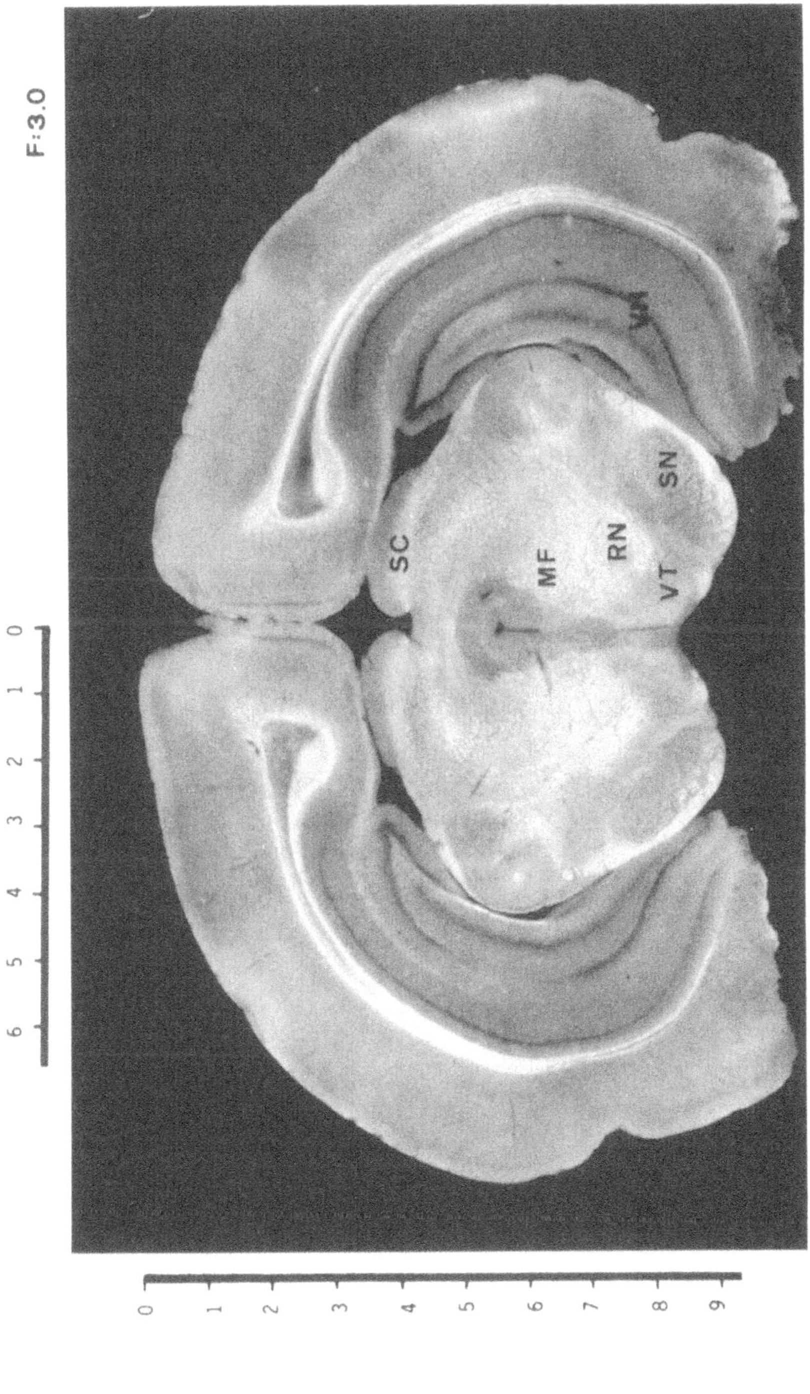

FIGURE 3.11B

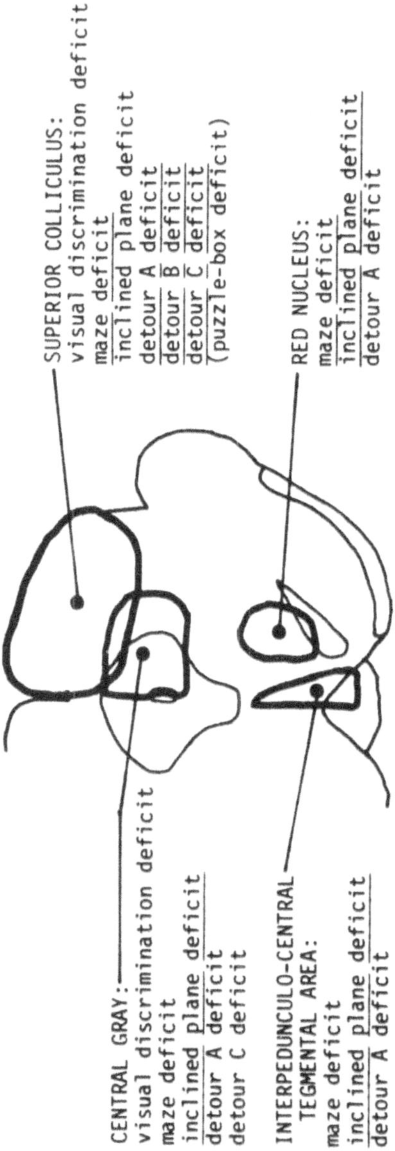
SUPERIOR COLLICULUS:
visual discrimination deficit
maze deficit
inclined plane deficit
detour A deficit
detour B deficit
detour C deficit
(puzzle-box deficit)
RED NUCLEUS:
maze deficit
inclined plane deficit
detour A deficit
CENTRAL GRAY:
visual discrimination deficit
maze deficit
inclined plane deficit
detour A deficit
detour C deficit
INTERPEDUNCULO-CENTRAL TEGMENTAL AREA:
maze deficit
inclined plane deficit
detour A deficit

FIGURE 3.12A

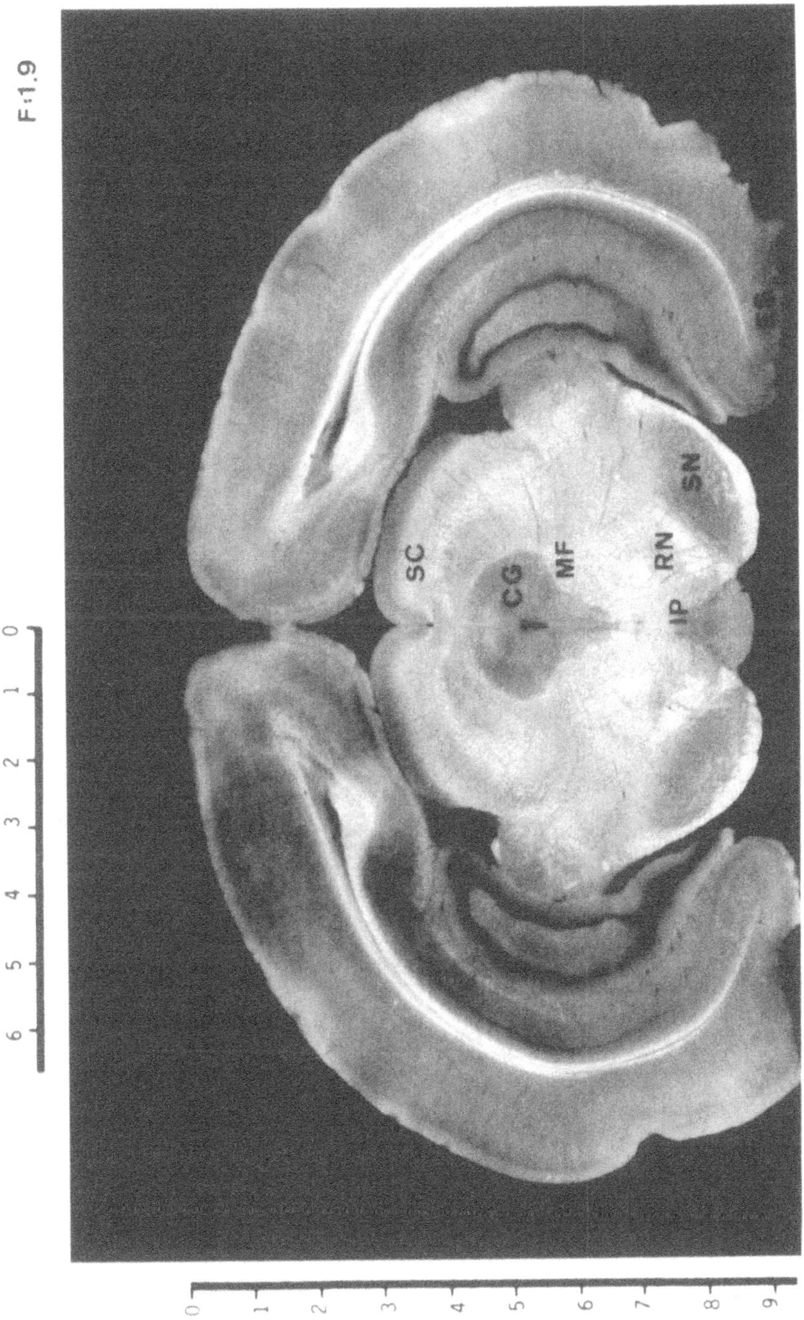

FIGURE 3.12B

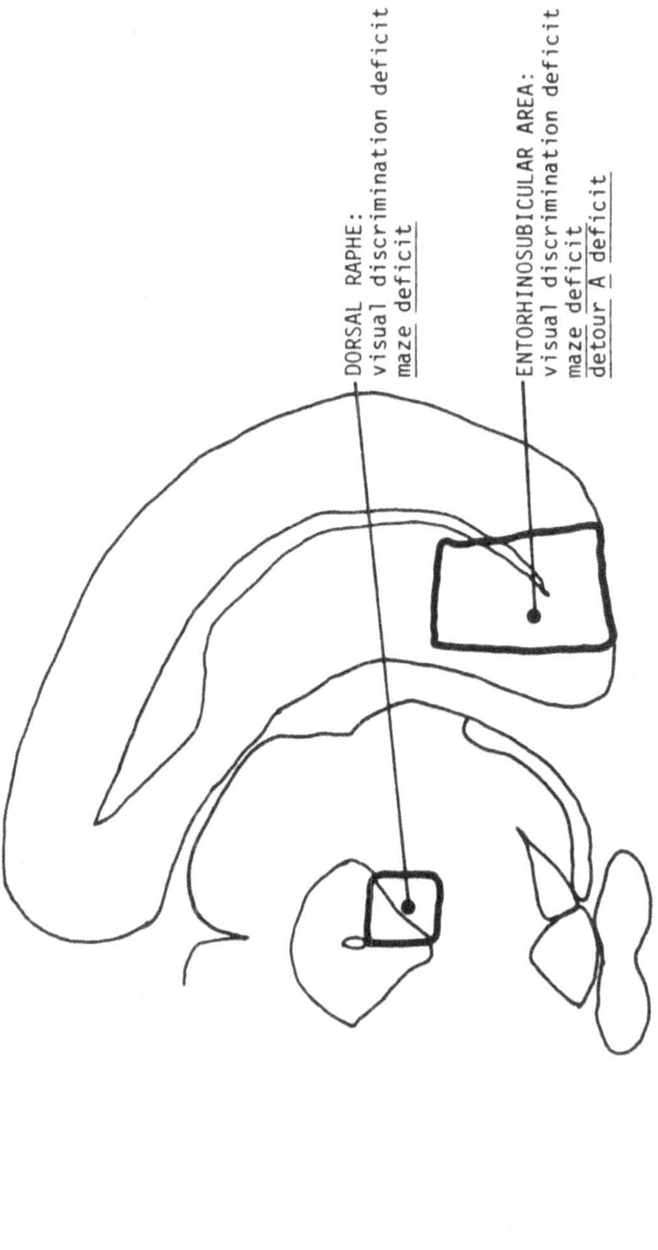
DORSAL RAPHE:
visual discrimination deficit
maze deficit
ENTORHINOSUBICULAR AREA:
visual discrimination deficit
maze deficit
detour A deficit

FIGURE 3.13A

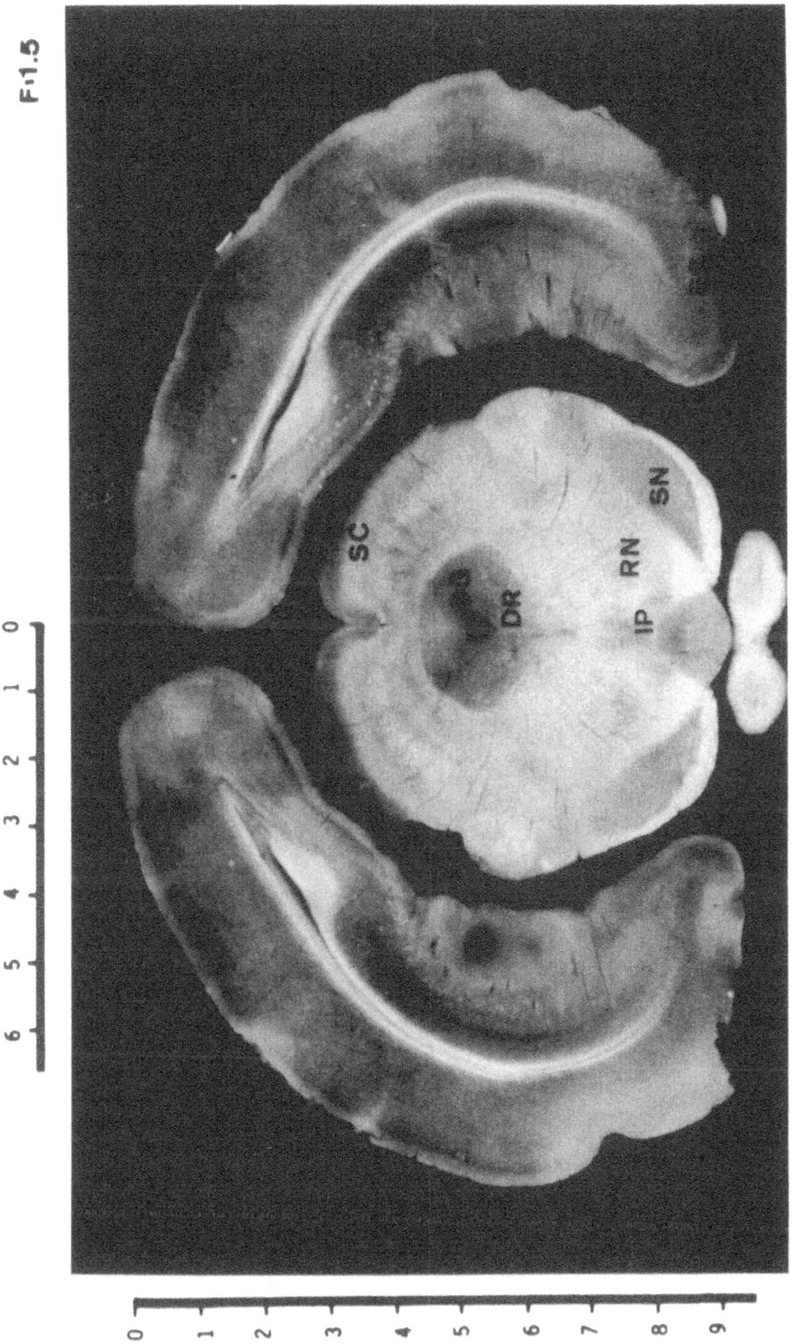

FIGURE 3.13B

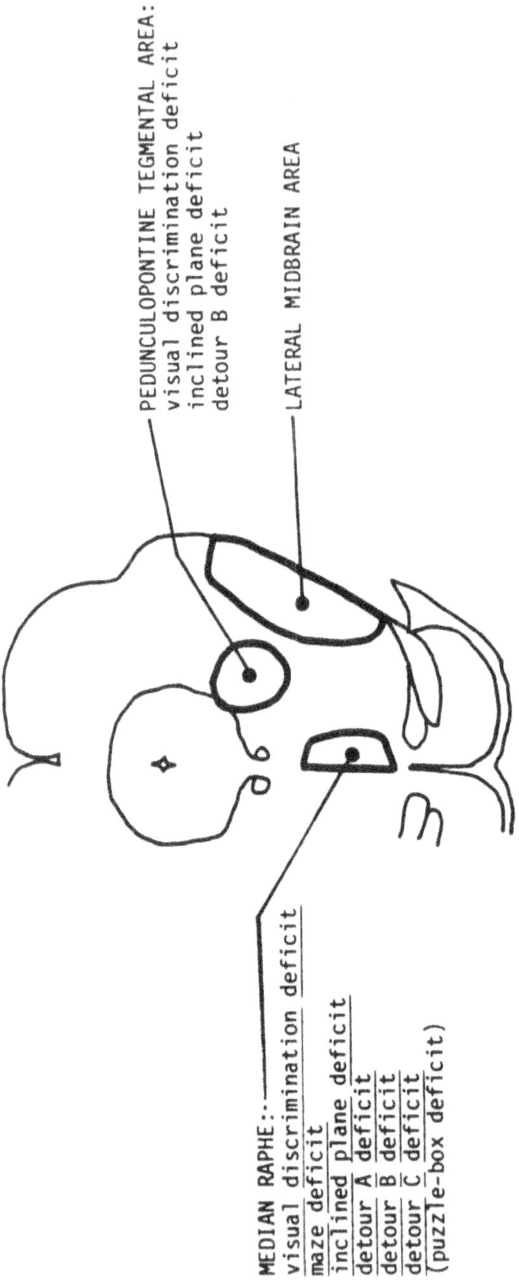
PEDUNCULOPONTINE TEGMENTAL AREA:
visual discrimination deficit
inclined plane deficit
detour B deficit
LATERAL MIDBRAIN AREA
MEDIAN RAPHE:
visual discrimination deficit
maze deficit
inclined plane deficit
detour A deficit
detour B deficit
detour C deficit
(puzzle-box deficit)

FIGURE 3.14A

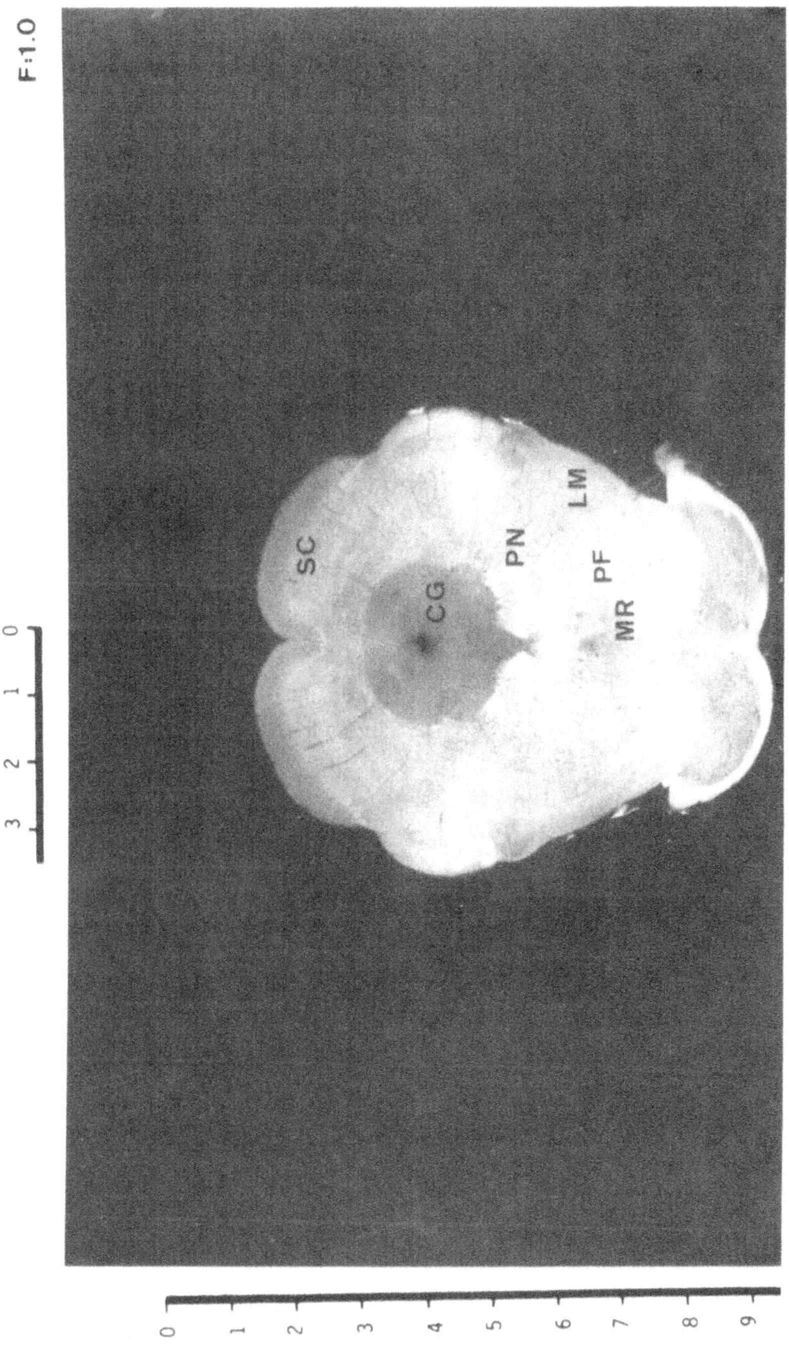

FIGURE 3.14B

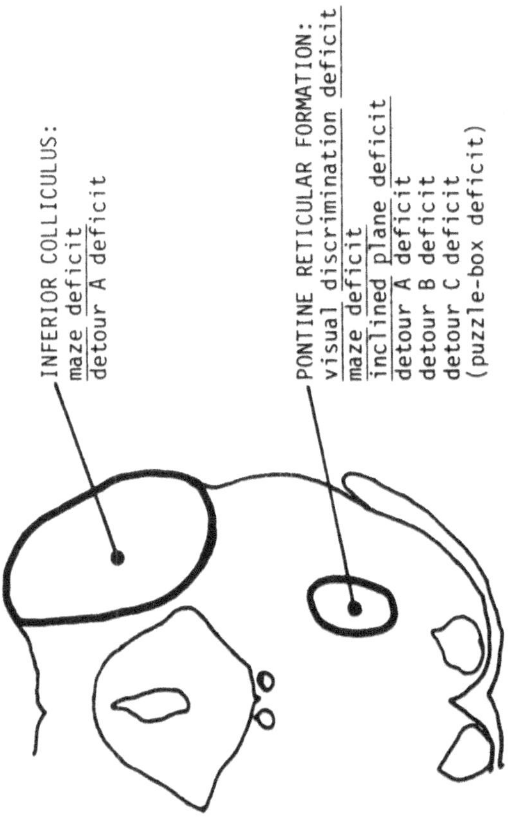
INFERIOR COLLICULUS:
maze deficit
detour A deficit
PONTINE RETICULAR FORMATION:
visual discrimination deficit
maze deficit
inclined plane deficit
detour A deficit
detour B deficit
detour C deficit
(puzzle-box deficit)

FIGURE 3.15A

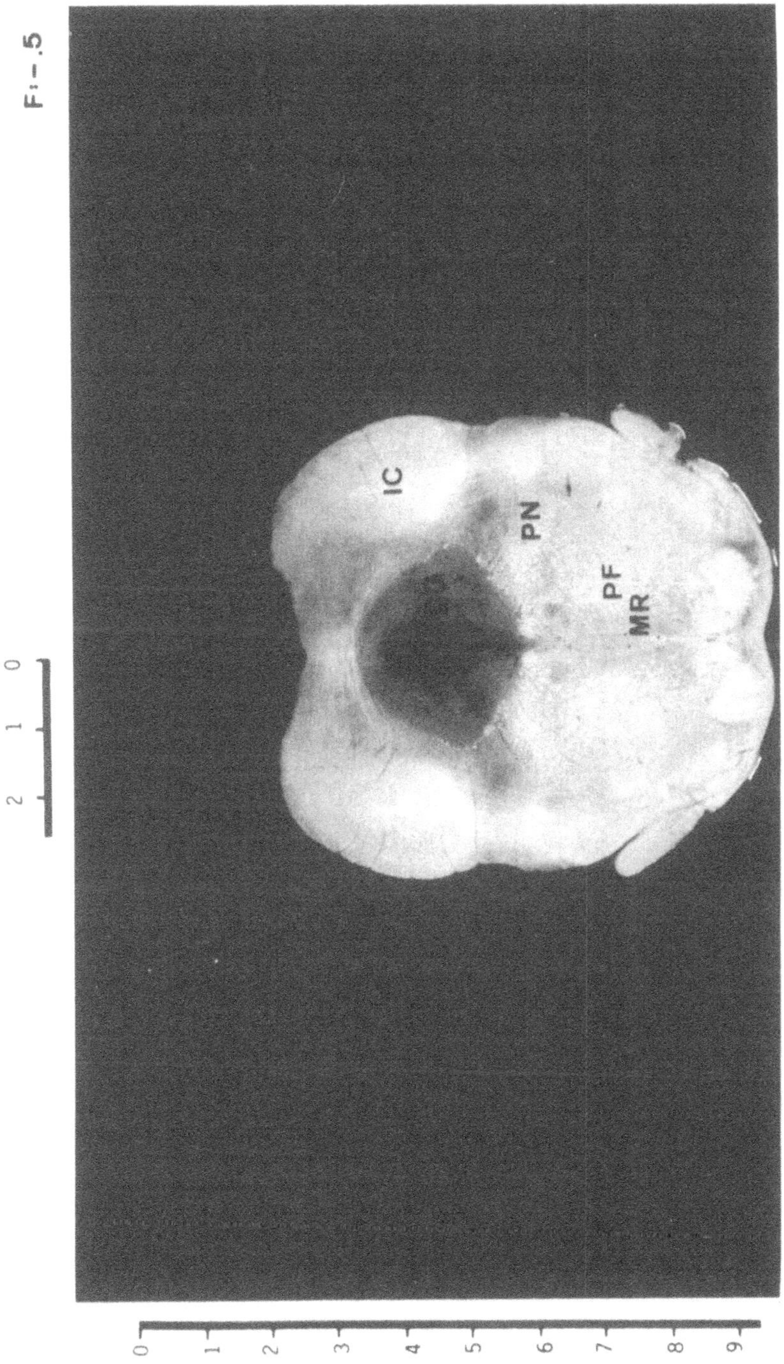

FIGURE 3.15B

4

Maps of Learning Systems

List of Learning Tasks and Corresponding Maps

Task	Map number
Visual discrimination	4.1
Inclined plane discrimination	4.2
Maze	4.3
Detour A	4.4

Map key

Moderate deficit 

Severe deficit 

Abbreviations

AC Frontocingulate cortex
AM Amygdala
AP Anterior pretectal nucleus
AT Anterior thalamus
CB Cerebellum
CG Central gray
CL Centrolateral thalamus
DC Dorsal caudatoputamen
DH Dorsal hippocampus
DR Dorsal raphe

E Eye
EP Entopeduncular nucleus
ES Entorhinosubicular area
FC Dorsal frontal cortex
GP Globus pallidus
H Habenula
HA Anterior hypothalamus
HV Ventromedial hypothalamus
IC Inferior colliculus
IP Interpedunculocentral tegmental area
LH Posterolateral hypothalamus
LM Lateral midbrain area
LT Lateral thalamus
MB Mammillary bodies
MD Mediodorsal thalamus
MF Rostromedial midbrain reticular formation
MR Median raphe
NA Nucleus accumbens septi
OC Occipitotemporal cortex
OL Olfactory bulb
P Parafascicular nucleus
PA Parietal cortex
PC Posterior cingulate cortex
PF Pontine reticular formation
PN Pedunculopontine area
PT Medial pretectal area
RC Rostral caudatoputamen
RN Red nucleus area
SC Superior colliculus
SF Septofornix area
SN Lateral substantia nigra
ST Subthalamus
VB Ventrobasal thalamus
VC Ventral caudatoputamen
VF Ventral frontal cortex
VH Ventral hippocampus
VL Ventrolateral thalamus
VM Ventromedial thalamus
VP Ventral pallidum
VT Ventral tegmental area

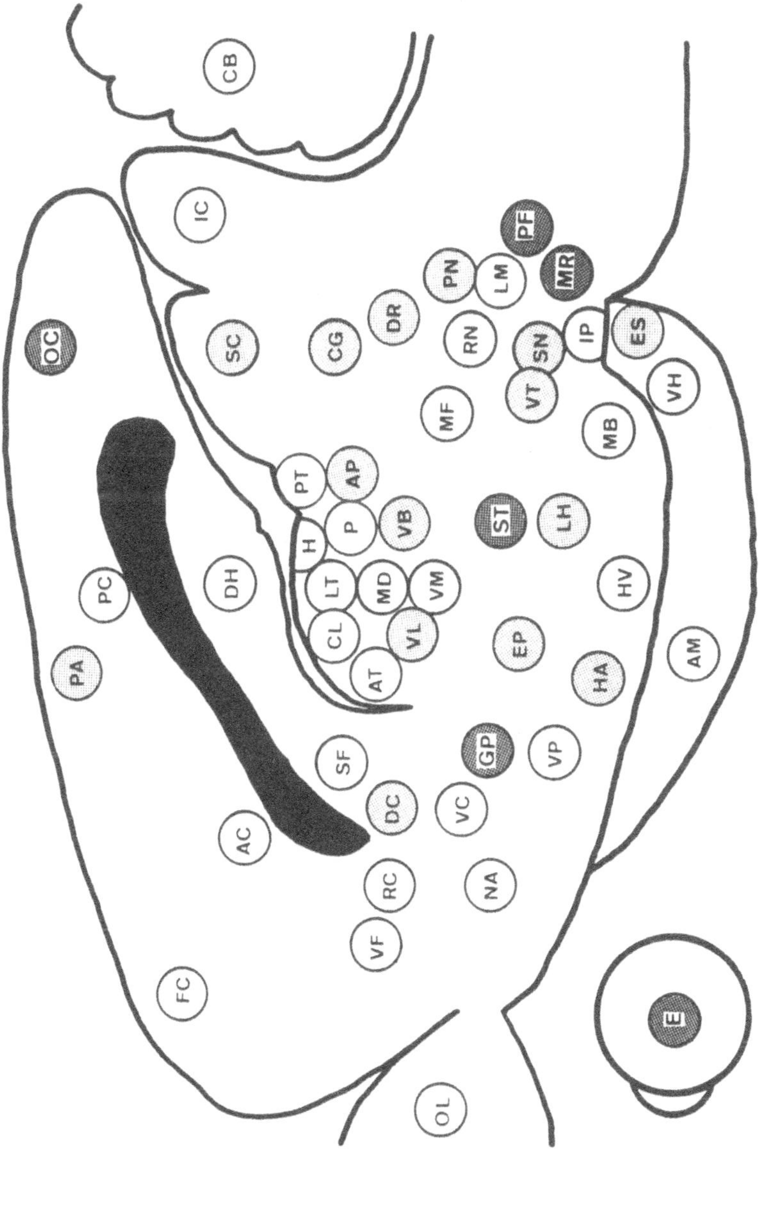

FIGURE 4.1. Visual discrimination learning system.

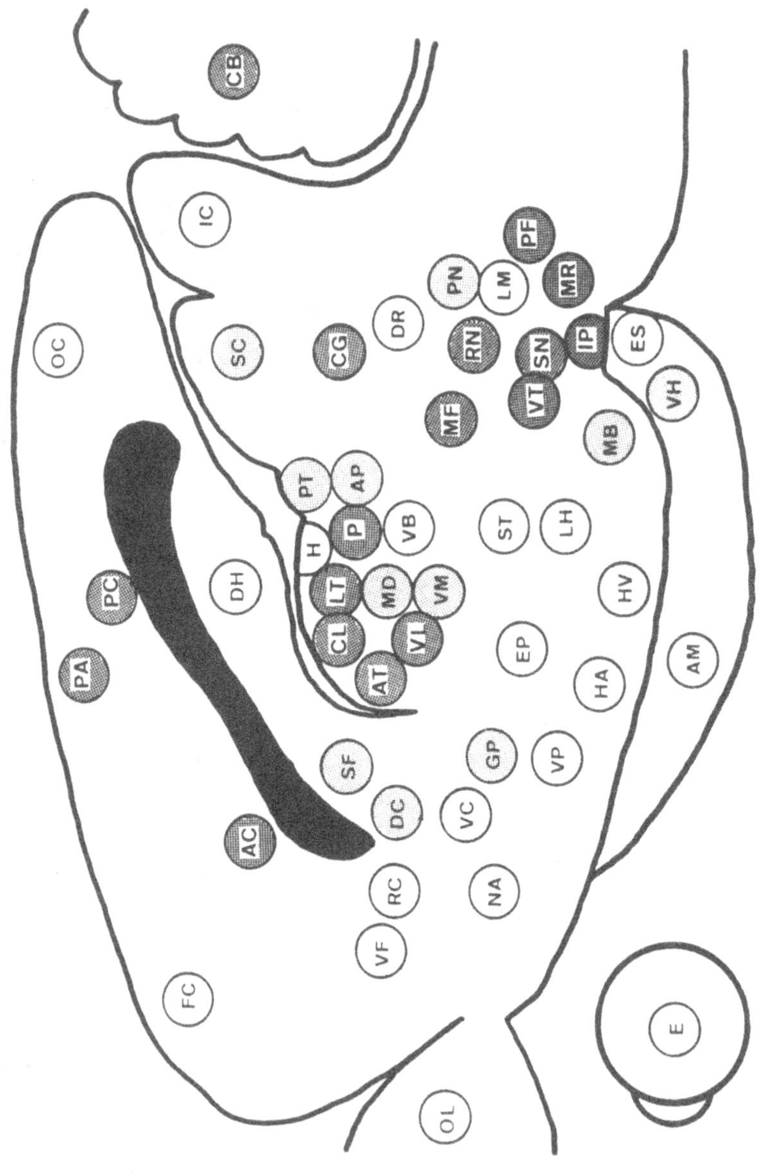

FIGURE 4.2. Inclined plane discrimination learning system.

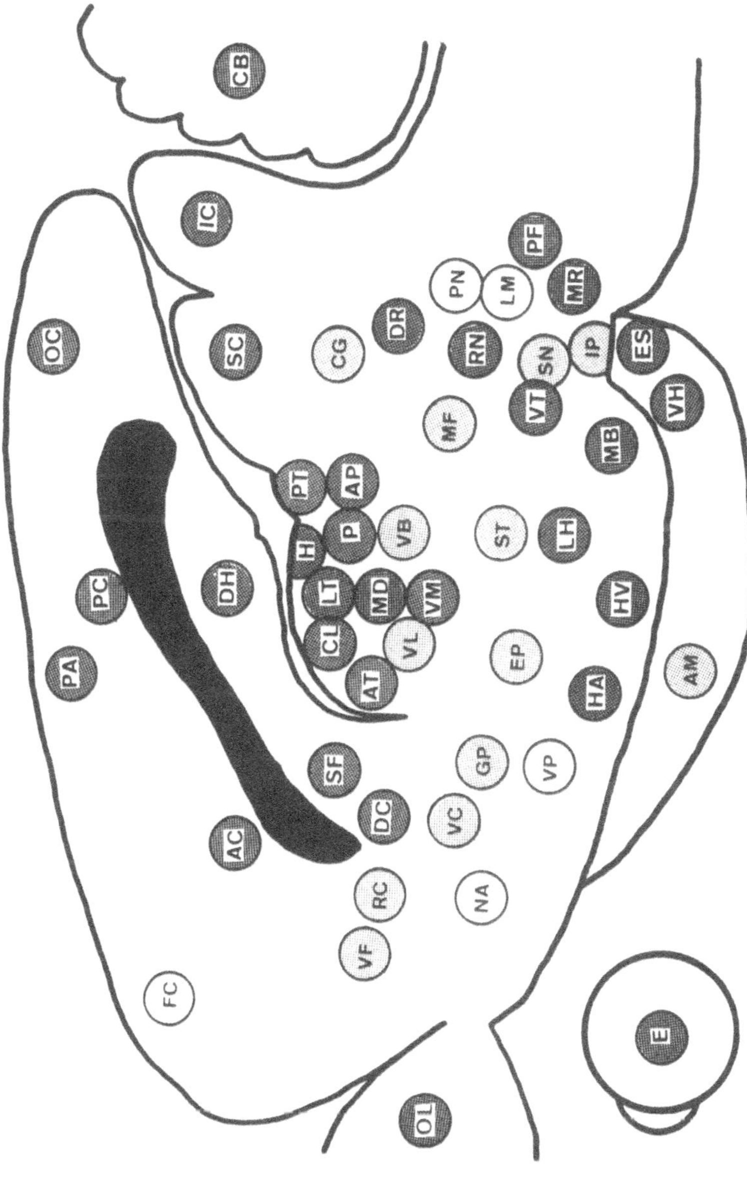

FIGURE 4.3. Maze learning system.

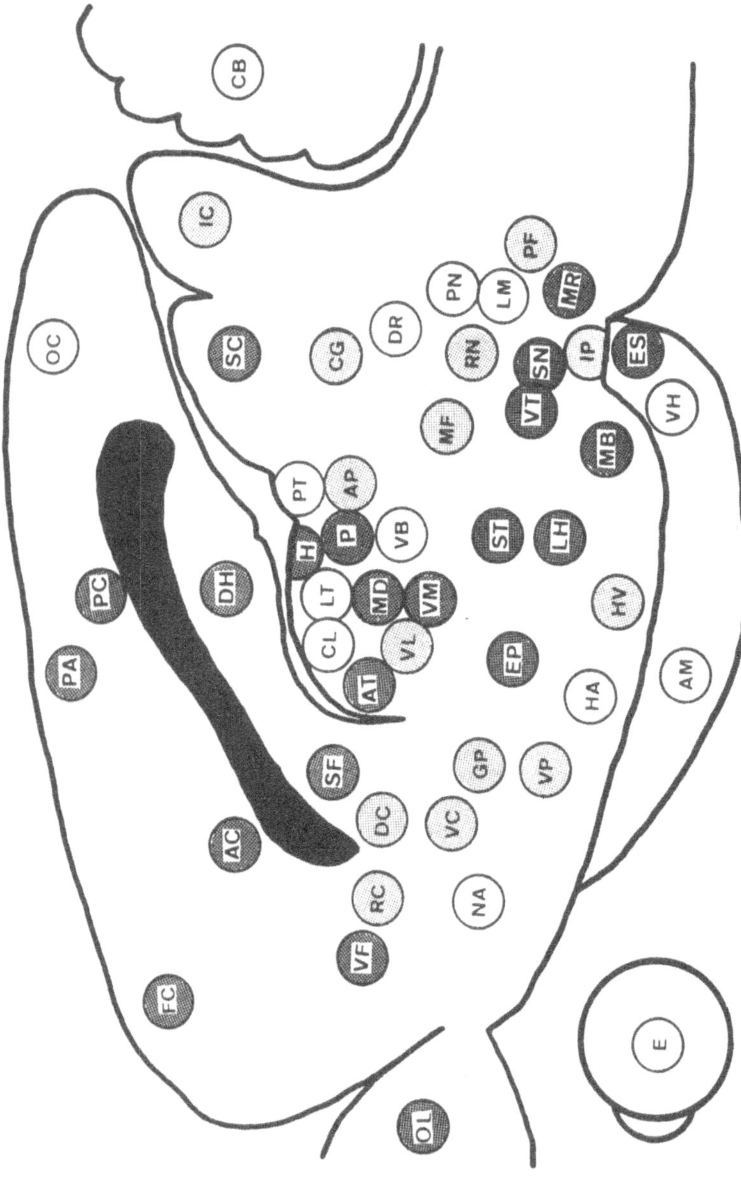

FIGURE 4.4. Climbing detour learning system.

5

General Findings

Throughout this research project, it has not been our intention to tease out the functions of discrete parts of the brain in learning. Rather, we have been primarily concerned with identifying ensembles of brain structures that emerge in importance for the acquisition of particular kinds of problem-solving tasks. In this respect, we agree with Luria (1966) that localizing complex mental functions (e.g., problem-solving) to a circumscribed area of the brain is unrealistic and that it is far better to envisage such functions as products of the activity of a "constellation" of interconnected nuclei situated at various levels of the central nervous system. Whereas Luria attempted to discover these constellations by studying the cognitive capacities of patients with damage to different parts of the brain, we chose to expose them by testing the learning ability of young rats with lesions to different brain sites.

But first, it would be worthwhile making some comments of a general nature about the learning syndromes assembled in Chapter 3 and the learning systems charted in Chapter 4.

Learning Syndromes

1. Inspection of the various syndromes suggests that the involvement of certain brain areas in learning may be more

complex and diffuse than currently conceptualized. Recent textbooks in physiological psychology, for example, typically discuss the occipitotemporal cortex, pretectal area, and superior colliculus only in relation to visual learning; the cerebellum, globus pallidus, and red nucleus only in relation to motor learning; and the cingulate cortex, amygdala, and median raphe only in relation to conditioned avoidance learning. Yet, the syndromes arising from lesions to these areas of the brain extended to classes of problem-solving tasks seemingly unrelated to visual, motor, or avoidance learning. One possible explanation of the diffuse complexity of many of the syndromes rests on the fact that most brain structures are not only traversed by extrinsic fiber systems, but are heterogeneous with respect to their intrinsic input and output pathways. Conceivably, extrinsic fiber systems and different subsets of intrinsic afferent and efferent projection pathways may take part in different functions and, as a consequence, participate in the acquisition of different problem-solving activities.

2. It appears that damage to virtually any part of the brain of the rat will provoke a learning deficit on at least one kind of problem-solving task. This conclusion is based on the finding that of the 50 brain sites canvassed with lesions, only two (nucleus accumbens and lateral midbrain area) were devoid of a learning syndrome. (The possibility exists that tests of conditioned avoidance learning might be sensitive to the presence of lesions to either the nucleus accumbens or lateral midbrain area—see Thompson, 1978b.)

3. The learning syndromes associated with lesions to the dorsal caudatoputamen, globus pallidus, ventrolateral thalamus, substantia nigra, ventral tegmental area, superior colliculus, median raphe, and pontine reticular formation are of particular interest because they are associated with learning deficits on all tasks composing the test battery (including the puzzle-box problems). These results suggest that this ensemble of structures may play a nonspecific role in learning, which will be fully discussed in Chapter 7.

4. The cerebral cortex and limbic forebrain (particularly the hippocampus) are regions of the mammalian brain most

frequently mentioned in connection with learning and memory. In other words, these two regions of the brain are generally believed to rank highest in "problem-solving complexity." Assuming that the number of learning deficits arising from a lesion to a given brain site is an index of the problem-solving complexity of that site, then the results reported in Chapter 3 have opened the way to a provisional examination of regional differences in problem-solving complexity. Obviously, conclusions based on any significant regional differences must be made with caution, especially in light of the fact that the relative number of learning deficits listed under a syndrome will probably be a function of such variables as the composition of the test battery, the length of the recovery period, the age and previous experience of the subjects, the size of the lesions, and the number of extrinsic fiber systems interrupted by the lesions.

In Table 5.1, the 50 different brain sites examined in this study are grouped under one of nine arbitrary divisions of the brain. The mean number of deficits is shown for each division (performance on the puzzle-box problems has been excluded from this analysis) together with the brain sites in each division associated with the highest and lowest number of deficits. According to this analysis, the limbic midbrain area, hypothalamus, thalamus, and basal ganglia exhibited the highest degree of problem-solving complexity, and the cerebral cortex and limbic forebrain were among the brain divisions exhibiting the lowest degree of problem-solving complexity.

This difference in problem-solving complexity between the cerebral cortex and limbic forebrain, on the one hand, and that associated with certain deep formations residing within the upper neuraxis, on the other, may be taken as evidence favoring the importance of the brainstem in acquiring adaptive, goal-directed responses, but other interpretations cannot be ruled out. For example, it might be argued that certain brainstem lesions abolish "energizing" influences on other cerebral areas (Luria, 1970), thus causing generalized disturbances in problem-solving behaviors. Or, in line with the notions of Berntson and Micco (1976), it might be argued that the broad range of

TABLE 5.1. *Regional Differences in the Total Number of Deficits (Problem-Solving Complexity)*

Brain division[a]	Number of brain sites	Mean	Highest	Lowest
Basal ganglia (DC,GP,EP,RC,SN,ST, VC,VP)	8	4.4[b]	DC,GP, and SN (6)	VC (2)
Limbic midbrain area (CG,DR,IP,MR,VT)	5	4.4[b]	MR and VT (6)	DR (2)
Thalamus (AP,AT,CL,LT,DM,P, VB,VL,VM)	9	4.1	VL (6)	VB (2)
Hypothalamus (AH,HV,LH,MB)	4	4.0	LH and MB (5)	AH (2)
Cerebral cortex (FC,OC,PA,VF)	4	3.5	PA (6)	VF (2)
Sensory pathways (E,OL)	2	3.5	OL (4)	E (3)
Brainstem (IC,LM,MF,PF,PN,PT, RN,SC)	8	3.3	PF AND SC (6)	LM (0)
Limbic forebrain (AC,AM,DH,ES,H,NA, PC,SF,VH)	9	2.9	AC,DH,PC, and SF (4)	NA (0)
Cerebellum (CB)	1	2		

[a]Abbreviations for individual brain sites are the same as those used in Chapter 4.
[b]Differed from the limbic forebrain area ($p < 0.05$, Mann–Whitney test).

problem-solving deficits arising from brainstem lesions is a reflection of the disorganization of those species-typical behavioral patterns that are involved in the expression of learned responses. After all, there are good reasons to believe that learning is a modification and elaboration of innate behavior patterns (Gallup, 1983) and that the process of learning is often innately guided (Gould & Marler, 1987).

5. The number of learning deficits associated with damage to a given structure cannot readily be explained in terms of the size of the lesions. The animals with cerebellar lesions, for example, suffered removal of a greater volume of brain tissue than those with frontal, parietal, occipitotemporal, or cingulate ablations, yet exhibited the fewest learning deficits. With respect to injuries to subcortical formations, those rats with lesions to the lateral midbrain area sustained at least twice the amount of damage to brainstem tissue in comparison to those with lesions to the ventral tegmental area or substantia nigra; nevertheless, the former failed to manifest a single learning deficit, while the latter manifested learning deficits on every problem composing the test battery (including the puzzle-box problems).

Learning Systems

1. The findings related to the charting of the four learning systems shown in Chapter 4 fall short of supporting either a strict localizationist position or a strict nonlocalizationist position with respect to the functions of the brain in problem-solving activities. On the one hand, the map derived from the study of the white–black discrimination problem (see Figure 4.1) revealed a reasonably localized ensemble of structures concerned with acquisition of visually guided behaviors—of the 50 brain sites investigated, less than 50% appeared to be critical for this class of learning. However, the map derived from the study of the maze habit (see Figure 4.3) disclosed what appears to be a relatively diffuse mechanism underlying the acquisition

of this class of problems—90% of the brain sites canvassed with lesions were associated with a maze learning deficit.

2. It appears that the complexity of the neural circuits underlying acquisition of any given problem-solving task is determined more by the intrinsic nature of the task than by its difficulty, as gauged by the number of trials or errors required for control subjects to reach a given learning criterion. This was suggested by the finding that the maze learning system (Figure 4.3) is appreciably more complex than the visual discrimination learning system (Figure 4.1), even though the maze habit was learned in fewer errors than the visual discrimination habit (about 10 errors as opposed to about 20 errors). Further support for this conclusion comes from the finding that the composition of the inclined plane discrimination learning system (Figure 4.2) is more extensive than that of the visual discrimination learning system, even though the former was learned in less than half of the errors associated with the learning of the latter.

3. The discovery that a number of different brain areas are essential for normal acquisition of a given problem-solving task helps to explain, at least in part, the almost ubiquitous appearance of recovery of function after brain damage (see Stein, Rosen, & Butters, 1974). Although several different explanations are available, a likely one is that functional recovery results from incomplete destruction of that system of neural structures concerned with the performance of a particular task. A lucid account of a similar position has been presented by LeVere (1975).

4. The illustrations presented in Chapter 4 disclose that no unique learning deficits arise from lesions to the neocortex. Any given learning deficit occurring with frontal, parietal, or occipitotemporal ablations was reproducible with discrete subcortical lesions. These findings underscore the importance of corticosubcortical and subcorticocortical interactions in learning and provide striking support for the argument that complex mental functions are not to be viewed exclusively as products of the interplay of various neocortical areas via corticocortical associations (Penfield, 1954a,b).

Dissecting the Learning Systems

By comparing the various lesion-defined learning systems mapped in Chapter 4, it has been possible to isolate several groups of brain sites (denoted in this monograph as "neural mechanisms") that appear to be recruited (emerge in importance) for the normal acquisition of particular kinds of problem-solving tasks. Some tasks, such as the visual and inclined plane discrimination habits, may require the engagement of only three of these neural mechanisms, whereas others (maze and detour problems) seem to involve the recruitment of at least five separate neural mechanisms.

Nonspecific Mechanism

This mechanism consists of the dorsal caudatoputamen, globus pallidus, ventrolateral thalamus, substantia nigra, ventral tegmental area, superior colliculus, median raphe, and pontine reticular formation (see Figure 5.1). Identification of this mechanism was made by determining which brain sites were associated with learning deficits on all problems, including the maze, two discrimination, three detour, and puzzle-box problems. Of the 50 brain sites examined with lesions, only those eight mentioned above were found to be the "common denominators" of the learning systems investigated to date. (While the parietal cortex, anterior pretectal nucleus, and midbrain central gray are included within the learning systems associated with the discrimination, maze, and detour problems, they are not involved in the learning system associated with puzzle-box tasks—see Thompson, Gallardo, & Yu, 1984a,b; Thompson, Huestis, & Yu, 1987; Thompson *et al.*, 1989a.)

This generalized problem-solving impairment associated with lesions to the nonspecific mechanism cannot readily be dismissed as being secondary to one or more behavioral disturbances commonly mentioned in connection with brain damage. For example, the nature of the test battery makes it

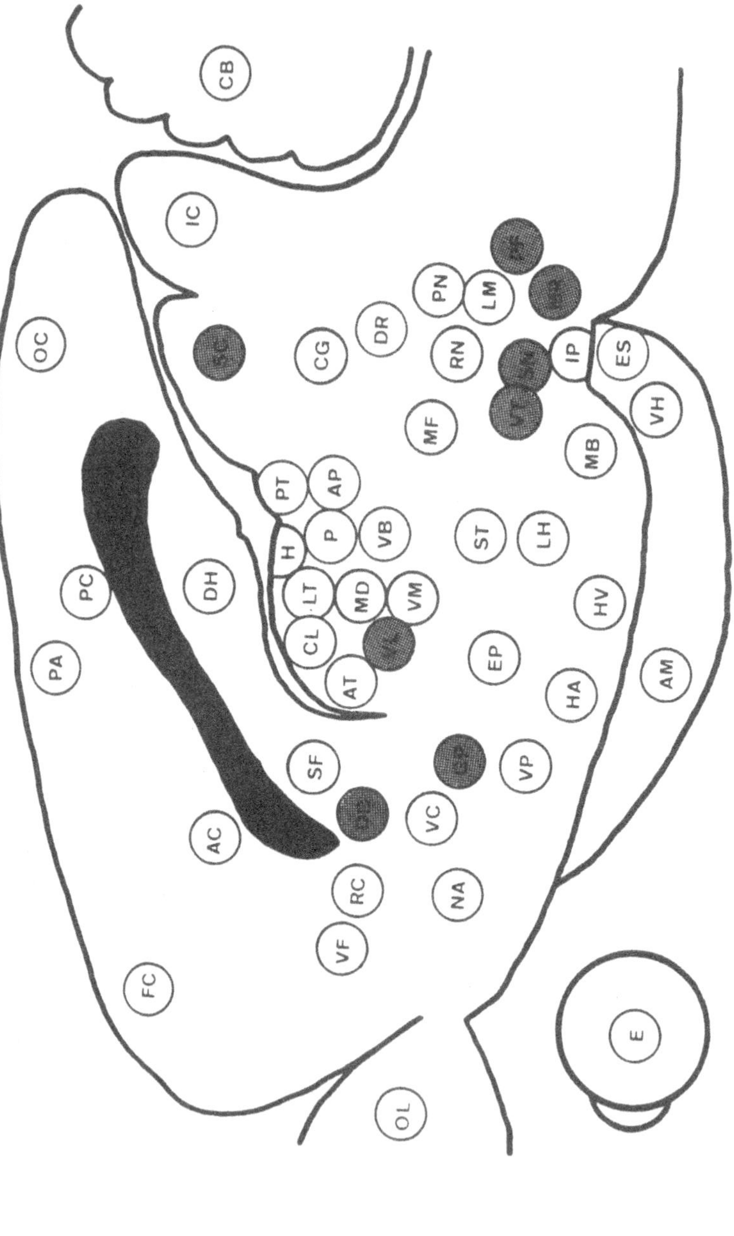

FIGURE 5.1. The eight structures (shaded areas) composing the "nonspecific mechanism."

unlikely that a specific sensory or motor defect underlies the appearance of a generalized learning impairment. Attributing this impairment to general malaise or to a defect in arousal or emotionality also seems unlikely, since most of the brain-damaged rats in question were healthy, alert, normally active, and emotionally stable at the time of the first learning test. (However, those groups with lesions to the median raphe or superior colliculus tended to be hyperactive relative to other brain-damaged and control rats.) With the possible exception of those animals with lesions to the ventral tegmental area or substantia nigra, a motivational involvement can be disregarded as a contributing factor since most of the brain-damaged rats with lesions to components of this nonspecific mechanism responded with what appeared to be normal vigor in pursuit (and in the presence) of the goal object. In the case of those rats with ventral tegmental lesions, the majority evidenced reduced motivation on both the appetitive and aversive tasks. Those rats with nigral lesions, on the other hand, manifested defective motivation on the appetitive tasks only. Parenthetically, a motivational involvement does not necessarily preclude the coexistence of a generalized problem-solving deficit (see Chapter 1).

This disclosure of a nonspecific mechanism presumably critical for overall problem-solving ability is subject to several general interpretations. First, it can be argued that the limited number of learning tasks explored in the present study is insufficient to infer that the eight identified structures serve a nonspecific function in problem-solving. For example, there is a conspicuous absence of data on olfactory, auditory, or tactile learning which rats unequivocally engage in. Tasks specifically measuring learning in a Skinner box situation or in a radial maze were also excluded from the test battery. It cannot be denied that our observations require further documentation, but it is doubtful that any additional tests of problem-solving ability will drastically alter the current composition of the nonspecific mechanism. Actually, the present study represents the third in a series of lesion surveys of the rat brain aimed at identifying nonspecific neural mechanisms in learning and memory (retention). The results of this third survey are remarkably similar to those reported in two previous surveys (Thompson, 1984), despite the use of weanling rats and detour problems. (In the two previous surveys, adult rats were used and the laboratory tasks involved not only discrimination and maze habits, but active avoidance responses and spatial reversal learning.)

Against this interpretation is the one that acknowledges that the eight brain regions under discussion are significant for general problem-solving, but views certain subgroups as having one function and other subgroups as having altogether different functions. For example, the caudatoputamen could be concerned with memory, the substantia nigra with motor control, the vertral tegmental area with motivation, and the pontine reticular formation with attention. Although not conclusive, the weight of the evidence seems to oppose this view. If memory disturbances can lead to a general problem-solving deficit, why did lesions to the mammillary bodies, hippocampus, or cerebral cortex fail to provoke a nonspecific learning impairment? Similarly, if a general problem-solving deficit can arise from abnormalities in motor control, why did destruction of the frontal (motor) cortex, cerebellum, or red nucleus impair acquisition of no more than half of the problems composing the test battery? And if a general problem-solving impairment can be reduced to a disturbance in motivation or attention, why did injuries to the hypothalamus, parietal cortex, or midbrain reticular formation fall short of producing learning impairments on all tasks composing the test battery?

The interpretation we favor envisages the nonspecific mechanism as an aggregate of anatomically interconnected neural regions whose combined activities are responsible for a unitary, though broad, function in problem-solving. That this may indeed be the case is suggested by the finding that virtually all components of the nonspecific mechanism are either elements of or anatomically related to the basal ganglia (Heimer, Alheid, & Zaborszky, 1985; Kitai, 1981; McGeer, McGeer, Itagaki, & Mizukawa, 1987). This prominent subcortical system has traditionally been viewed as having a broad unitary function in motor control, but has more recently been proposed to play an additional role in cognitive activities, including sequencing behaviors (Albert, 1978; Buchwald, Hull, Levine, & Villablanca, 1975; Cummings, 1986), attention (Cummings, 1986; Hassler, 1977), cognitive sets (Buchwald *et al.*, 1975), learning and memory (Hassler, 1980; Phillips & Carr, 1987), and thought processes (Carlsson, 1988; Nauta & Feirtag, 1986). In view of the foregoing considerations, it is reasonable to propose that the basal ganglia constitute the core of the nonspecific mechanism.

The foregoing conceptualization that the basal ganglia and anatomically related nuclei represent a sort of global problem-solving system is admittedly open to question because certain brain regions composing (or intimately connected to) the basal ganglia (e.g., entopeduncular nucleus, ventromedial thalamic nucleus, subthalamus, and pedunculopontine tegmental nucleus) are absent from the nonspecific mechanism. These negative findings, however, are not overwhelming and may reflect either the existence of a specialized subset of basal gangliar nuclei essential for general problem-solving behavior or the presence of redundancy within this anatomical system.

Visuospatial Attentional Mechanism

Like those structures composing the nonspecific mechanism, the parietal cortex, anterior pretectal nucleus (formerly referred to in the rat as the nucleus posterior thalami), and mesencephalic central gray are included within the learning systems associated with the discrimination, maze, and detour habits. However, they differ from the former insofar as they are not implicated in the acquisition of latch-box or puzzle-box tasks (Thompson *et al.*, 1984a, 1987, 1989a) and, consequently, cannot be assumed to play a nonspecific role in learning. Since these three structures are critical for the acquisition of habits involving orientation (and navigation) to objects in extrapersonal space, they are tentatively linked to a mechanism concerned with attention to visuospatial cues (Figure 5.2). (Conceivably, this mechanism may sustain a function that Semmes,

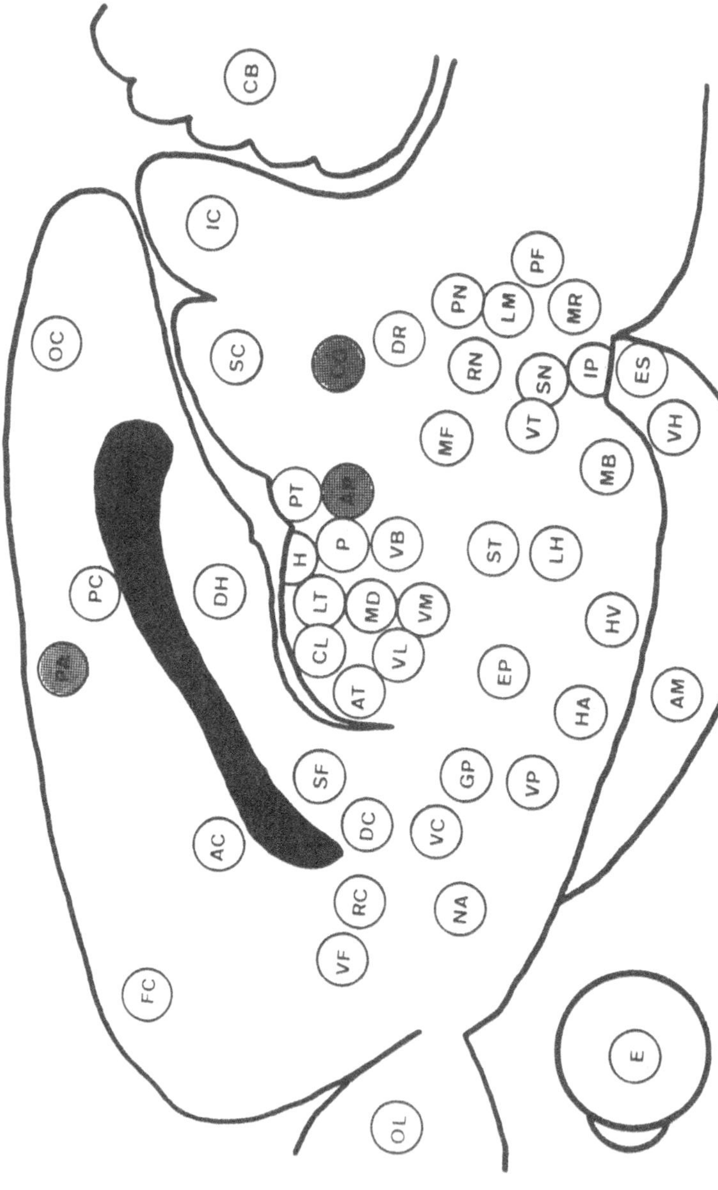

FIGURE 5.2. The three structures (shaded areas) composing the "visuospatial attentional mechanism."

1967, referred to as a "general spatial factor.") Young rats with selective lesions to the parietal cortex, anterior pretectal nucleus, and central gray are generally healthy, alert, motivated, and free from any obvious motor or emotional disorders.

From an anatomical point of view, these three brain regions could conceivably serve a common function. For example, these regions share the feature of being targets of either direct or indirect somatosensory afferents (Andrezik & Beitz, 1985; Beitz, 1987; Tracey, 1985). Furthermore, descending projections from the parietal cortex have been traced to both the anterior pretectal nucleus (Roger & Cadusseau, 1984; Wise & Jones, 1977) and midbrain central gray (Meller & Dennis, 1986). Similarly, the central gray appears to send a contingent of fibers to the region of the anterior pretectal nucleus (Eberhart, Morrell, Krieger, & Pfaff, 1985), while the latter in turn sends a projection to the parietal cortex (Donoghue, Kerman, & Ebner, 1979; Nothias, Peschanski, & Besson, 1988). Most of these interconnecting pathways, however, are weak and require further validation.

Admittedly, viewing this ensemble of brain regions as a visuospatial attentional mechanism is in the domain of sheer speculation. While the parietal cortex has been implicated in both visuospatial and attentional mechanisms (Heilman, Watson, Valenstein, & Goldberg, 1987), the anterior pretectal nucleus and central gray have not. Since these two brainstem structures have frequently been discussed in relation to nociception (Kelly, 1985), it can be argued that animals with injuries to either structure alone would be handicapped in learning virtually any task based on aversive motivation. Yet, lesions to either the anterior pretectal nucleus or central gray have produced learning (or retention) deficits in situations in which nociceptive stimulation was excluded (Storozhuk, Ivanova, & Tal'nov, 1985; Thompson & Spiliotis, 1981; Detour Problem A of present study). These findings clearly underscore the importance of gathering further information on the possible functions of the anterior pretectal nucleus and midbrain central gray in various problem-solving situations.

Visual Discrimination Mechanism

This mechanism consists of nine structures, namely, the eye (retina), occipitotemporal cortex, entorhino-subicular area, ventrobasal thalamus, entopeduncular nucleus, subthalamus, anterior hypothalamus, posterolateral hypothalamus, and dorsal raphe (see Figure 5.3). Identification of this mechanism was easily made by determining which structures included within the visual discrimination learning system were not included within the nonvisual inclined plane (vestibular–proprioceptive–kinesthetic) discrimination learning system. This comparison seemed reasonable since both habits, although similar with respect to motivational conditions and motor demands, differed mainly in terms of the sense modality engaged in the discrimination task. Although not investigated in the present study, the optic pathways from the retina to the lateral geniculate nuclei of the thalamus are likely components of this visual discrimination mechanism as well (see Thompson, 1976).

The composition of this specific mechanism presumably essential for normal acquisition of conventional two-choice visual (but not nonvisual) discrimination habits is surprising, to say the least. Obviously, the core structures of this mechanism consist of the retinogeniculooccipitotemporal pathway. Conceivably, some of the remaining components of this mechanism gain their functional significance in visual discrimination learning because of their connections with one or more of these core structures. This is clearly the case with respect to the subthalamus (zona incerta) and anterior hypothalamus (of which the suprachiasmatic nucleus is a part) to the extent that the former receives projections from the occipitotemporal cortex (Shammah-Lagnado, Negrao, & Ricardo, 1985) and ventral lateral geniculate nucleus (Legg, 1979; Ribak & Peters, 1975), while the latter receives projections from the retina (Kita & Omura, 1982) and ventral lateral geniculate nucleus (Legg, 1979; Ribak & Peters, 1975). The posterolateral hypothalamus and entorhinosubicular area may represent similar cases insofar as the former has been reported to be the target of projections from the retina (Kita & Omura, 1982) and the latter may

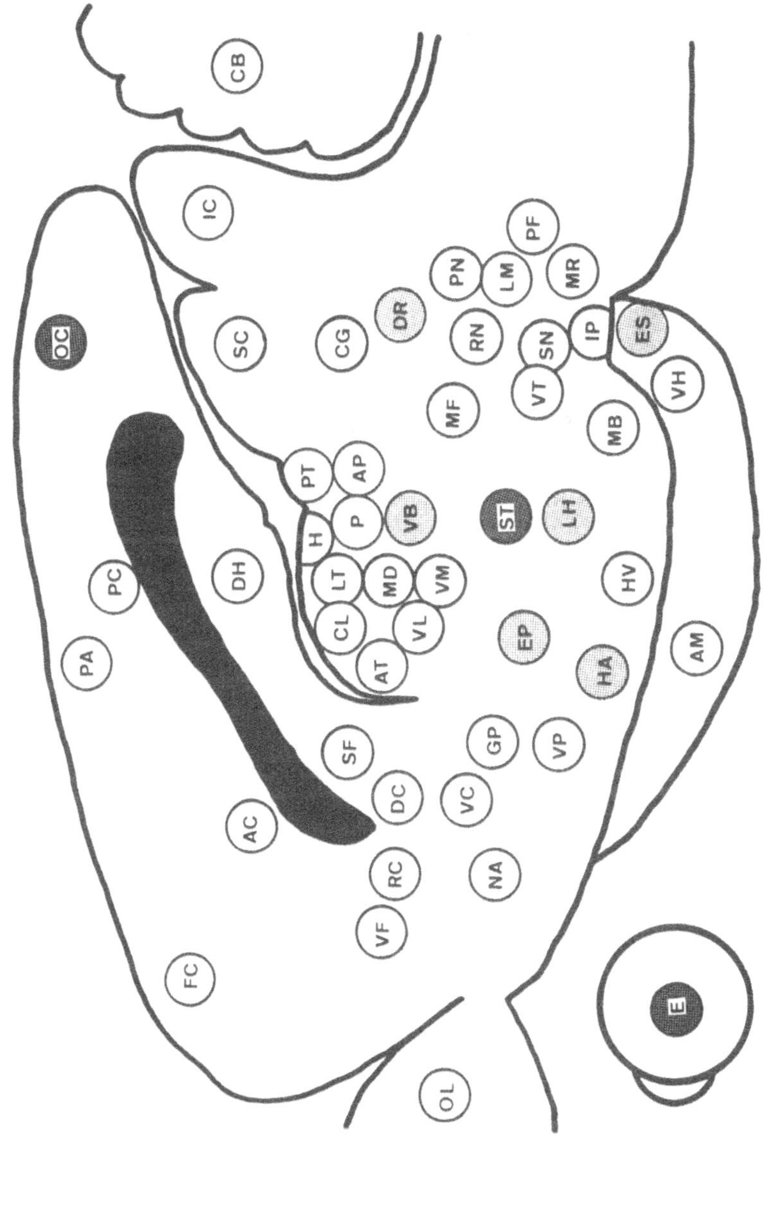

FIGURE 5.3. The nine structures composing the "visual discrimination mechanism." (See Chapter 4 for significance of differential shading.)

be the recipient of efferent fibers from the occipitotemporal cortex (Bayer, 1985). With respect to the dorsal raphe, although it does not receive efferent projections from the retinogeniculooccipitotemporal pathway, it does send a projection to the lateral geniculate nuclei (Pasquier & Villar, 1982; Villar, Vitale, Hokfelt, & Verhofstad, 1988). This leaves for consideration the remaining two structures of this specific mechanism (entopeduncular nucleus and ventrobasal thalamus), which may not have direct connections with the retinogeniculooccipitotemporal pathway. Although other explanations are available—for example, we may have erred in the inclusion (Type I error) or exclusion (Type II error) of certain brain sites within this mechanism because of the use of small samples—the possibility exists that these two structures may function in some fashion as intermediary links between the visual discrimination mechanism and either the nonspecific mechanism or the visuospatial attentional mechanism. After all, acquisition of the black–white discrimination task can be conceived as requiring coactivation of the visual discrimination mechanism, the visuospatial attentional mechanism, and the nonspecific mechanism.

Vestibular–Proprioceptive–Kinesthetic Discrimination Mechanism

This is the most extensive specific mechanism (see Figure 5.4) uncovered in the current study, consisting of 16 structures (frontocingulate cortex, posterior cingulate cortex, septofornix area, ventral hippocampus, anterior thalamus, centrolateral thalamus, lateral thalamus, mediodorsal thalamus, ventromedial thalamus, parafascicular nucleus of the thalamus, medial pretectal area, mammillary bodies, rostral midbrain reticular formation, interpedunculocentral tegmental area, red nucleus, and cerebellum). It was identified by determining those structures included within the inclined plane discrimination learning system which were not included within the visual discrimination learning system. Additional components of this mechanism may include the vestibular nuclei of the medulla, the medial longitudinal fasciculus, the medial lemniscus, and

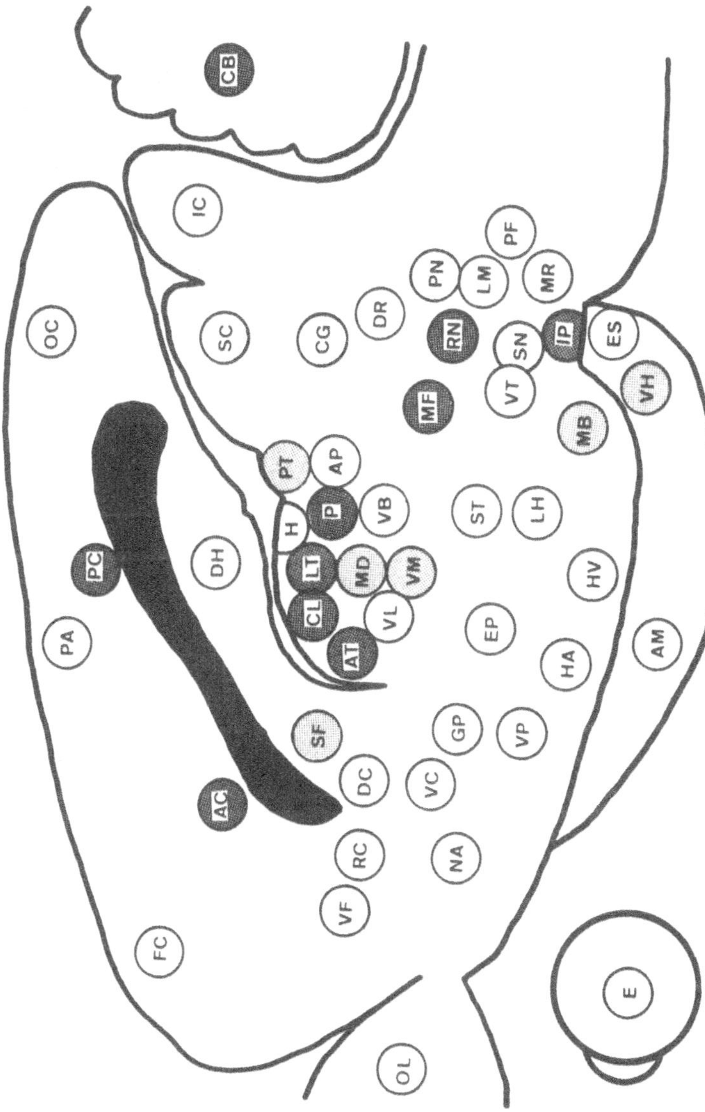

FIGURE 5.4. The 16 structures composing the "vestibular–proprioceptive–kinesthetic discrimination mechanism." (See Chapter 4 for significance of differential shading.)

certain other ascending spinal or brainstem pathways transmitting relevant vestibular, proprioceptive, or kinesthetic information (Mehler & Rubertone, 1985). (Conspicuously absent from this mechanism is the ventrobasal thalamus; conceivably, our lesions to this thalamic complex may have been too small to elicit a somatosensory deficit sufficient to impair acquisition of the inclined plane problem—see Frommer, 1981.) The extensiveness of this mechanism is not likely due to chance, since a remarkably similar ensemble of brain sites was recognized earlier in connection with lesion studies on retention of the inclined plane discrimination habit in adult rats (Thompson, Arabie, & Sisk, 1976).

From an anatomical point of view, this mechanism appears to consist of two relatively distinct "core networks," one involving cerebellorubrothalamic circuitry (Faull & Carman, 1978; Haroian, Massopust, & Young, 1981; Snider, Maiti, & Snider, 1976) and the other involving mammillothalamocortical circuitry (Jones, 1985; Papez, 1937). The components of the former would consist of the cerebellum, red nucleus, interpedunculocentral tegmental area, rostral midbrain reticular formation, medial pretectal area, parafascicular nucleus, ventromedial thalamic nucleus, and centrolateral thalamic nucleus, while those of the latter would consist of the mammillary bodies, lateral thalamus, mediodorsal thalamus, anterior thalamic nuclei, septofornix, ventral hippocampus, frontocingulate cortex, and posterior cingulate cortex.

What is equally remarkable about this vestibular–proprioceptive–kinesthetic discrimination mechanism is that the two core networks appear to serve altogether different functions. The cerebellorubrothalamic circuit has traditionally been assumed to be responsible for those motor functions concerned with the coordination of movements and the performance of skilled acts (Evarts & Thach, 1969). The functions attributed to the mammillothalamocortical circuit, on the other hand, have ranged from the mediation of emotional experience (Papez, 1937) to memory processing (Lhermitte & Signoret, 1976). In light of these considerations, an explanation must be sought for the finding that the normal acquisition of a problem-solving task involving the discrimination of vestibular–proprioceptive–kinesthetic cues (as opposed to normal acquisition of a problem-solving task involving the discrimination of visual cues) requires the integrity of two seemingly distinct pathways, one concerned with motor functions and the other concerned with emotional and memory functions.

In our view, a reasonable explanation of these phenomena can be had by assuming that (1) the acquisition of the inclined plane discrimination, the performance of skilled acts, the experience of emotion, and the formation of certain kinds of memory all require the processing of vestibular, proprioceptive, and/or kinesthetic signals, and (2) the processing of vestibular–proprioceptive–kinesthetic signals depends on the combined activities of those structures composing the mammillothalamocortical and cerebellorubrothalamic networks. The first assumption receives support from the findings that skilled motor acts clearly have a vestibular–proprioceptive–kinesthetic basis (Konorski, 1967; Luria, 1966); that peripheral bodily changes, which would result in vestibular, proprioceptive, and kinesthetic feedback, can be viewed as determinants as well as correlates of emotional experience (Izard, 1971); and, in the case of the rat, that memory processing functions associated with spatial tasks, such as spontaneous alternation, left–right discriminations, position reversals, and complex mazes, appear to be mediated, at least in part, by vestibular–proprioceptive–kinesthetic information (Douglas, Clark, Hubbard, & Wright, 1979; Thompson, Hale, & Bernard, 1980; Watson, 1907; Zoladek & Roberts, 1978). (Hocherman, Aharonson, Medalion and Hocherman, 1988, have recently demonstrated in humans that proprioceptive inputs are involved in the perception of the immediate extrapersonal space.) While compelling evidence favoring the second assumption is lacking, it does receive some support from the observations that certain structures within each network (cerebellum, medial pretectal area, red nucleus, centrolateral thalamus, mediodorsal thalamus, and lateral thalamus) are targets of somatosensory (proprioceptive–kinesthetic) and/or vestibular input (Giesler, Menetrey, & Basbaum, 1979; Mackel & Noda, 1989; Mehler & Rubertone, 1985; Nagata,

1986; Tracey, 1985), while others may be recipients of such impulses via relays in the locus ceruleus (Loughlin & Fallon, 1985), cerebellum (Haroian *et al.*, 1981), reticular formation (Zemlan, Leonard, & Pfaff, 1978), and pretectal complex (Robertson, Kaitz, & Robards, 1980), to mention only a few. Collateral support comes from the findings that selective lesions to most components of the mammillothalamocortical and cerebellorubrothalamic networks not only disrupt performance of the inclined plane discrimination problem, but disrupt performance of learned manipulative acts (Thompson, Gates, & Gross, 1979), mazes (Thompson, 1974), and active avoidance (fear) conditioned responses (Thompson, 1978b) as well.

Clearly, the foregoing conceptualization of the vestibular–proprioceptive–kinesthetic discrimination mechanism is not without its faults. For example, there appear to be few major anatomical links interconnecting the cerebellorubrothalamic network with the mammillothalamocortical network. Furthermore, lesions to certain components of the vestibular–proprioceptive–kinesthetic discrimination mechanism (e.g., anterior and lateral thalamic nuclei and posterior cingulate cortex) have not been found to retard acquisition of learned manipulative acts (Thompson *et al.*, 1984a,b). These faults, however, may not be very serious. Concerning the first one, the functional coupling of the two seemingly unrelated neural networks could be achieved through the mediation of either the nonspecific mechanism or the visuospatial attentional mechanism (see Snider & Maiti, 1976). With respect to the second, focal lesions within the vestibular–proprioceptive–kinesthetic discrimination mechanism may not invariably give rise to a given expected learning deficit owing to the presence of parallel circuits that could bypass certain disconnections in primary circuits.

Finally, if it is assumed that the vestibular–proprioceptive–kinesthetic discrimination mechanism in the human brain is similar in composition to that in the rat brain, then the identification of this mechanism has enormous implications for the notion that learning disabilities and dyslexia have a cerebellar–

vestibular basis (Frank & Levinson, 1973; Kohen-Raz, 1986). Although these disabilities in human patients have traditionally been assumed to be due primarily to abnormalities at the level of the cerebral cortex (see, for example, Pavlidis & Fisher, 1986), there is evidence that the vast majority of these patients manifest cerebellar–vestibular disorders (see Levinson, 1988, for a recent review of this literature). These findings have led to the conceptualization that learning disabilities and dyslexia are not of a primary cortical origin, but rather of a cerebellar–vestibular origin, and that any observed dysfunctions within the neocortical domain may be secondary to impaired cerebellar and brainstem processes (Levinson, 1988). This theory is all the more plausible now that the possibility exists that cerebellar–vestibular dysfunction would directly impact on the activities of such brain structures making up the vestibular–proprioceptive–kinesthetic mechanism as the cingulate cortex, most nuclei of the thalamus, and midbrain reticular formation. Parenthetically, the proposal that there may be a cerebellar–vestibular basis for fears, phobias, and related anxiety states (Levinson, 1986) is compatible with the finding that the vestibular–proprioceptive–kinesthetic discrimination mechanism contains elements of the limbic forebrain.

Place-Learning Mechanism

This mechanism consists of nine structures, namely, the olfactory bulbs, ventral frontal cortex, dorsal hippocampus, amygdala, habenula, rostral caudatoputamen, ventral caudatoputamen, ventromedial hypothalamus, and inferior colliculus (Figure 5.5). This neural mechanism was exposed by determining which structures included within the maze learning system were not included within either the visual or the inclined plane discrimination learning system. Choosing to implicate this mechanism with place learning was based primarily on the O'Keefe and Nadal (1978) distinction between guidance learning and place learning. With respect to guidance learning, the correct (or incorrect) route to the goal is marked by a conspic-

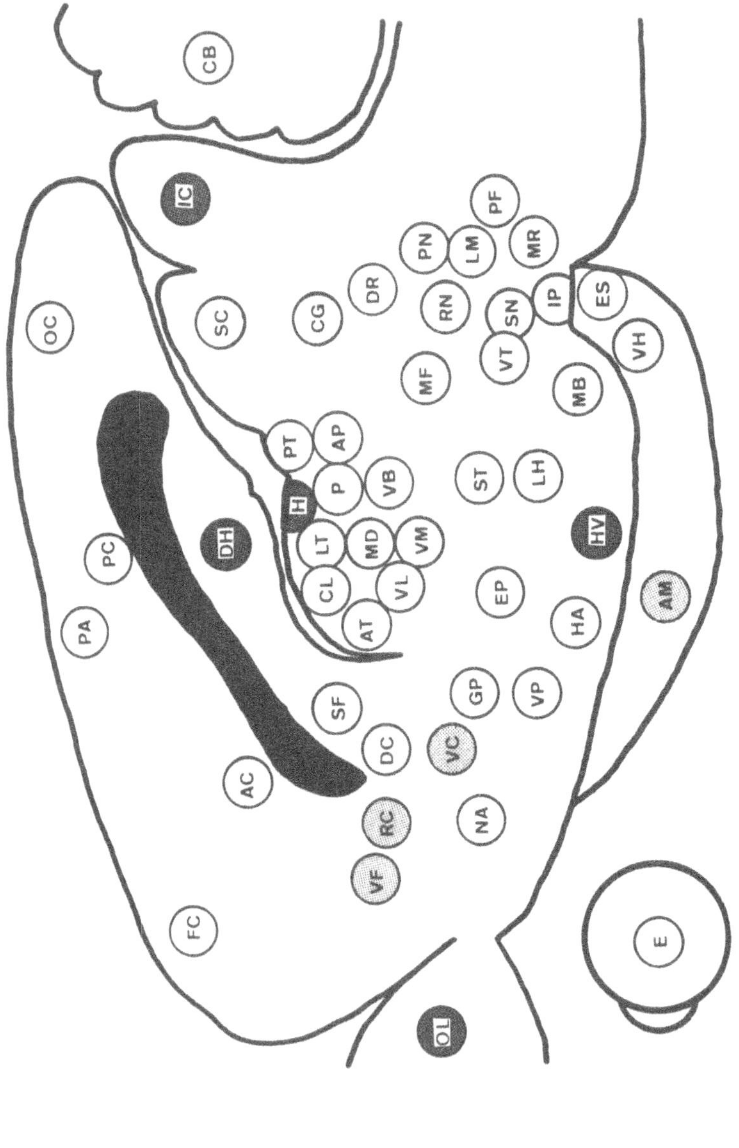

FIGURE 5.5. The nine structures composing the "place-learning mechanism." (See Chapter 4 for significance of differential shading.)

uous cue, such as a light. Most conventional two-choice sensory discrimination tasks, such as the white–black and inclined plane discrimination problems investigated in the current study, would engage this kind of learning. Place learning, on the other hand, consists of the development of an internal representation of the spatial layout of the environment. Most complex maze tasks are assumed to be dependent on place learning. The fact that the hippocampus is essential for place learning in particular and "cognitive mapping" in general (O'Keefe & Nadel, 1978), together with the fact that our place-learning mechanism includes the dorsal hippocampus, constitutes further support for the proposed designation.

Although the importance of the hippocampus in acquiring a complex maze is not in doubt, this formation must work in concert with other neural structures to mediate the complexities of place learning. Conceivably, the remaining eight brain structures composing the place-learning mechanism may form the major part of a matrix within which the hippocampus processes spatial and nonspatial cues in the construction (and utilization) of cognitive maps. This conceptualization is reasonably defensible in light of the findings that all of these structures, except the inferior colliculus, entertain afferent or efferent connections with the hippocampal formation (Herkenham & Nauta, 1979; Nauta & Domesick, 1981; Swanson & Kohler, 1986; Switzer, de Olmos, & Heimer, 1985; Wyss, Swanson, & Cowan, 1979).

It should be emphasized, however, that other brain regions, such as the occipitotemporal cortex, septofornix, and mamillary bodies, may also play a role in place learning even though these regions may be associated with learning of certain types of discrimination (guidance) tasks. There is evidence, for example, that the occipitotemporal cortex may have a nonvisual function not only in rats but in cats and monkeys as well and that this nonvisual function may have something to do with cognitive mapping (Lashley, 1943; Lubar, Schostal, & Perachio, 1967; Orbach, 1959; Thompson, 1982b). Conceivably, the septofornix and mamillary bodies may likewise have a "nonsensory" function that contributes to the production and utilization of cognitive maps.

Response Flexibility Mechanism

A response flexibility mechanism was identifiable by determining which brain structures included within the detour learning system were not included within either the visual discrimination, inclined plane discrimination, or maze learning system. Linking this mechanism to response flexibility was based on the premise that the detour problem is ideally suited to study the ability of any given experimental subject to discard unsuccessful responses (usually dominant among the alternative response strategies) and to try novel ones, until the correct one is found (see Thompson *et al.*, 1984). The discrimination and maze habits can hardly be viewed as satisfactory tests of response flexibility, not only because the number of alternative responses is generally limited to one (approach the left alley rather than the right one), but because the alternative response has a high probability of being performed once the prepotent response is consistently unreinforced.

Only two structures emerged from this analysis, namely, the dorsal frontal cortex and ventral pallidum (Figure 5.6). Because of the relatively small number of components involved, the existence of a response flexibility mechanism consisting of at least the dorsal frontal cortex and ventral pallidum must be viewed with caution. Another reason for exercising restraint is that it cannot be assessed a priori whether it is the task difference or the motivational difference that is responsible for the centrality of the dorsal frontal cortex and ventral pallidum in the acquisition of detour problems—it should be remembered that the maze and discrimination problems were acquired under aversive motivation, while the detour problems were acquired under appetitive motivation.

On the other hand, there are two compelling reasons for at least tentatively accepting the existence of the currently constituted response flexibility mechanism. First, the ventral pallidum (portions of which are occupied by large cholinesterase-reactive cells that form a component of the nucleus basalis) is a source of afferent fibers to the dorsal frontal cortex (Bigl, Woolf, & Butcher, 1982; Johnston, McKinney, & Coyle, 1981; Wenk,

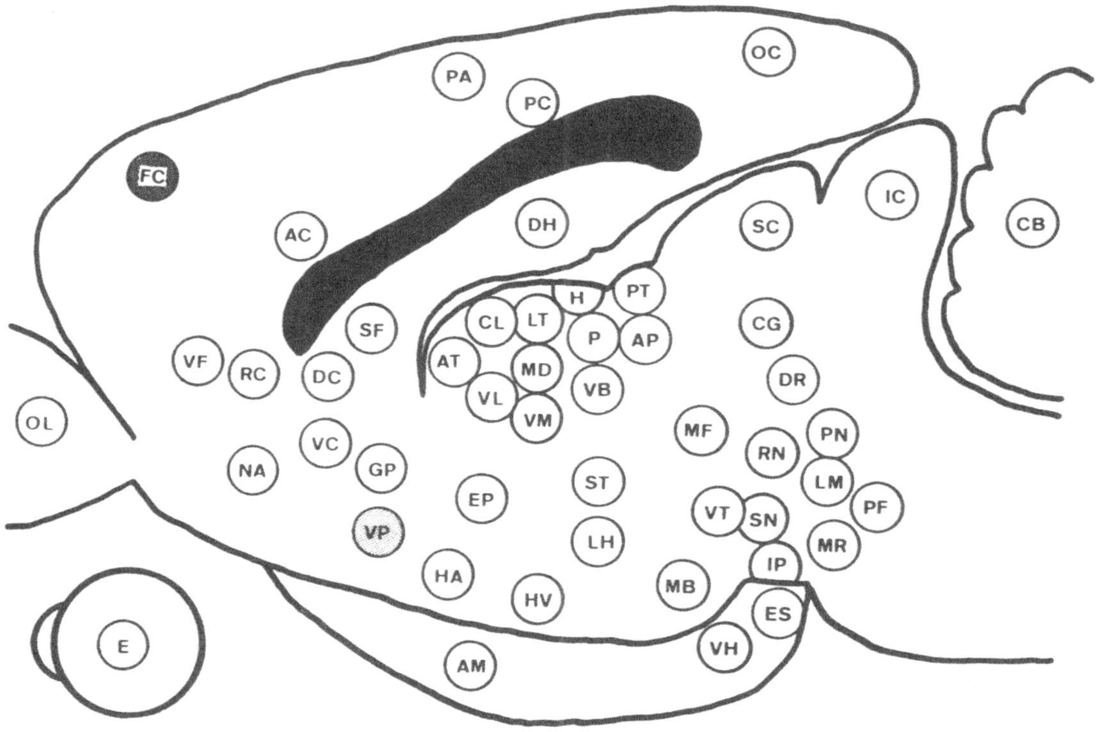

FIGURE 5.6. The two structures composing the "response flexibility mechanism." (See Chapter 4 for significance of differential shading.)

Cribbs, & McCall, 1984). Second, the dorsal frontal cortex and ventral pallidum are functionally related to the extent that lesions of either region impair not only retention of latch-box (Thompson *et al.*, 1979) tasks and active and passive avoidance responses (Thompson, 1978b), but acquisition of spatial reversals (Thompson, 1983) and a tunnel-digging detour problem as well (unpublished findings). Of course, the extent to which response flexibility is required for the performance of these tasks remains to be explicated.

Summary

Of the 50 brain sites canvassed with lesions, the foregoing six identified mechanisms have encompassed all but three, two of which (nucleus accumbens and lateral midbrain area) were not associated with a single learning deficit. The remaining brain site (pedunculopontine tegmental nucleus) was unique in the sense that it was the only structure associated with learning deficits restricted to the visual and inclined plane discrimination tasks.

It should be noted that these neural mechanisms may not be confined to the class of laboratory tasks from which they were derived. The visual discrimination mechanism, for example, is represented within the maze learning system—lesions to any given component of this mechanism were found to produce a significant learning deficit on the maze habit. Interestingly, every component of the vestibular–proprioceptive–kinesthetic discrimination mechanism is also represented within the maze learning system. These findings suggest that the maze habit, besides involving place learning, includes visual and vestibular–proprioceptive–kinesthetic learning as well. (In all probability, the relationship among these three types of learning derives from the fact that place learning is contingent on the processing of visual and vestibular–proprioceptive–kinesthetic cues.)

The detour problem, like the maze habit, can also be viewed as being mediated by a compound of separate neural mechanisms. For example, besides the response flexibility mechanism, it involves the place learning mechanism—of the nine structures composing the place learning mechanism, eight are included within the detour learning system. It is not entirely clear, however, whether the detour problem taps vestibular–proprioceptive–kinesthetic discrimination learning to the same degree as visual discrimination learning. Inspection of the detour learning system reveals that 69% of the components of the vestibular–proprioceptive–kinesthetic discrimination mechanism are represented as compared to only 44% of the components of the visual discrimination mechanism.

Based on the foregoing analysis, it is now possible to characterize the learning systems shown in Chapter 4 in terms of their component neural mechanisms. As shown in Table 5.2, normal acquisition of all laboratory tasks investigated in the current project was found to require the integrity of the nonspecific mechanism. Whether or not a second (visuospatial attentional), third (visual discrimination), fourth (vestibular–proprioceptive–kinesthetic discrimination), fifth (place learning), or sixth (response flexibility) mechanism is required for normal acquisition depends on the nature of the task, including its cognitive, sensory–perceptual, motor, and motivational features. Thus, the visual and nonvisual discrimination problems are alike in sharing two neural mechanisms, but differ with respect to the third mechanism. Interestingly, the maze appears to be a task that involves visual discrimination and vestibular–proprioceptive–kinesthetic discrimination learning; however, the maze differs from sensory discrimination habits in that a place-learning mechanism must also be engaged. Inspection of Table 5.2 further reveals that the detour problem likewise involves place learning and may depend to some degree on the integrity of the vestibular–proprioceptive–kinesthetic discrimination and visual discrimination mechanisms. On the other hand, the detour problem differs from the maze and discrimination problems by virtue of its dependence on a response flexibility mechanism.

There are at least three other ways to look at these mechanisms. First, from a psychometric point of view, the nonspecific mechanism could represent a "general factor" or "*g*" because this mechanism emerges in importance in all problem-solving

TABLE 5.2. *Neural Mechanisms Critical for Acquisition of Various Laboratory Tasks*[a]

Neural mechanism	White–black discrimination	Maze	Inclined plane discrimination	Detour
	Laboratory task			
Nonspecific	X	X	X	X
Visuospatial attentional	X	X	X	X
Visual discrimination	X	X		X
Vestibular–proprioceptive–kinesthetic discrimination		X	X	X
Place learning		X		X
Response flexibility				X

[a]The latch-box/puzzle-box task has been excluded from this table because of incomplete data; however, this task requires the nonspecific mechanism but not the visuospatial attentional mechanism.

tasks studied to date. The remaining five mechanisms, in contrast, would represent "group factors" or "abilities" which participate in certain kinds of problem-solving tasks, but not others. Second, from a cognitive perspective, the nonspecific mechanism bears a resemblance to what Fodor (1983) terms "central systems," which are assumed to mediate global cognitive processes involving problem-solving, thought, reasoning, and the planning of intelligent actions, to mention a few. The five remaining neural mechanisms identified in this study, on the other hand, could be likened to the corresponding "input systems" or "input modules" which take the products of one or more sensory analyzing structures, transform them into usable compounds, and deliver them to the central system, which, in turn, assembles them into wholes necessary for the mediation of global cognitive processes. And finally, from a neurological standpoint, the nonspecific mechanism could constitute Penfield's (1958) "centrencephalic system," which is responsible for integrating and coordinating the cognitive activities of specific mechanisms of the brain among which the visuospatial attentional, visual discrimination, vestibular–proprioceptive–kinesthetic discrimination, place learning, and response flexibility mechanisms could be a part. These conceptualizations are dealt with in greater detail in the remaining chapters of this book.

The foregoing conceptualizations, which imply that different problem-solving tasks require the recruitment of both common and separate neural mechanisms for their solution, are remarkably similar to a recently developed hypothesis about the cognitive operations of the human brain (Posner, Peterson, Fox, & Raichle, 1988). Based on new data from neural imaging studies of word reading, it was hypothesized that many elementary operations, each selectively localized within the brain, are involved in any cognitive task and that the task itself is not performed by any one brain region. However, "A set of distributed brain areas must be orchestrated in the performance of even simple cognitive tasks" (Posner *et al.*, 1988, p. 1627). In our view, the nonspecific mechanism is responsible for such orchestration (see Chapter 7).

6

The Nonspecific Mechanism and Psychometric *g*

In preceding chapters, evidence has been presented to support the existence of a nonspecific mechanism concerned with problem-solving in the white rat. The central feature of these findings has been the fact that certain lesions are capable of impairing performance on a wide variety of tasks.

Up to now, we have discussed the neuroanatomical correlates of problem-solving behavior from a general psychological perspective. Indeed, our studies over the past decade have invariably been approached from the *general–experimental* standpoint, i.e., a search for neural processes that mediate problem-solving behavior in all members of a given species. This model is traditional in experimental neuropsychology, where mean scores of lesioned groups are usually contrasted with scores of controls, using analysis of variance or some similar statistical technique. One of the unfortunate by-products of a correlative neuroanatomy based solely on such an approach, especially multiple studies confined to the investigation of a limited number of structures, has been an excessive emphasis on a "modular" view of brain functioning (Fodor, 1983).

Throughout the history of psychology, a parallel tradition, called *differential psychology,* has evolved, characterized by a search for features of behavior that are most salient for distinguishing members of a given species (usually human) from one another. Periodically, calls for greater integration of experimental and differential approaches in the design of experiments have surfaced (e.g., Carroll, 1988; Cronbach, 1957, 1975; Eysenck, 1961, 1982a; Keating, 1984; Sternberg, 1985), but there is little evidence that such suggestions have been accorded much more than lip service.

Our own results have brought us to a confluence of the experimental and differential traditions. Because of the unique questions posed by our findings, it was deemed necessary to employ the methods of differential psychology, now most often associated with psychometrics. As might be expected from the preceding chapters, our overriding consideration in this area has become the relationship between the nonspecific mechanism in the white rat, which we have elsewhere referred to as the "general learning system" (Thompson *et al.*, 1986), and that elusive construct known as "intelligence." In order to fully explore this relationship, several questions have to be answered, the most salient of which are: (1) Could a psychometric *g* (traditionally interpreted as general intelligence) be extracted from a varied set of problem-solving performances by the white rat? (2) Would specific factors, or group factors, distinctive from *g*, be found in the same set of performances? (3) Would there be lesions that produce deficits in psychometric *g*, while other lesions would not? And (4) would such lesions correspond to those structures identified as constituting the nonspecific mechanism?

Analysis of the Control Group

As a first step in answering these questions, the data from 75 sham-operated (i.e., control) animals were considered. Beginning at approximately 42 days of age, the animals were first run on the three detour problems and then on the black–white discrimination, three-cul maze, and inclined plane discrimination. As mentioned elsewhere, the detour problems were run under a food and water incentive, while the remaining three tasks involved escape/avoidance of foot shock. The experimen-

TABLE 6.1. Test Battery for 75 Sham-Operated Animals

Test	Measure	Motive
Detour A	Errors	Food/water
Detour B	Errors	Food/water
Detour C	Errors	Food/water
Response latency (detours)	Seconds	Food/water
Black–white discrimination	Errors	Escape/avoid shock
Three-cul maze	Errors	Escape/avoid shock
Inclined plane	Errors	Escape/avoid shock

tal paradigms, motives, and resulting performance measures are shown in Table 6.1.

Before analyzing the results, the error scores from the three detour tests were collapsed into a single measure. Even though the paradigmatic variations from which these scores were derived satisfied the criterion of "conceivable independence" (Carroll, 1988) for dependent measures, the fact that the same problem-box was used in each variation (only the barriers were altered) could have produced an "instrument factor" in any later factor analysis. Furthermore, inspection of the data clearly confirmed the fact that there was positive transfer of learning from one detour problem to the next.

The legitimacy of this concern for independence is further indicated by the Pearson product–moment correlations among the total error scores for the three detour tests (see Table 6.2). It is obvious that performances on the detour problems are interdependent, with an average intercorrelation among the three tests of 0.492 ($p < 0.01$). This itself is not an unusual finding, as similarly high levels of intercorrelation have been found among other complex problems sharing paradigmatic likeness (Lashley, 1929; Thorndike, 1935; Tryon, 1931).

Next, scores on all five dependent variables were intercorrelated. The resulting correlation matrix is shown in Table 6.3. (Response latency, an index of appetitive motivation, refers to

TABLE 6.2. Pearson Product–Moment Correlation Coefficients for Three Detour Tests[a]

	Detour		
Test	A	B	C
Detour A	1.00		
Detour B	0.58	1.00	
Detour C	0.45	0.44	1.00

[a]For $N = 75$, with three variables, the following significance levels apply: $p < 0.05$, $r = 0.286$; $p < 0.01$, $r = 0.351$.

TABLE 6.3. Correlation Matrix for 75 Sham-Operated Animals on Five Measures[a]

Measure	1	2	3	4	5
1. Detour (combined errors)	1.00				
2. Black-white discrimination	−0.04	1.00			
3. Maze	0.08	−0.09	1.00		
4. Inclined plane	0.01	0.04	0.08	1.00	
5. Response latency (pooled time scores)	0.32	−0.04	0.16	−0.13	1.00

[a]For $N = 75$, with five variables, the following significance levels apply: $p < 0.05$, $r = 0.304$; $p < 0.01$, $r = 0.362$.

the time elapsing from the moment the start box door is opened to the moment the rat enters the choice chamber of the detour box.) From the correlation matrix, it is apparent that there is, for the most part, only *minor intraindividual consistency* in performance from one task to another. The only significant correlation ($r = +0.32$; $p < 0.05$), that between the summed detour error measure and response latencies in the detour tests, probably represents paradigmatic, rather than true, intraindividual consistency; i.e., an animal more sure of the problem solution will tend to leave the start box more rapidly. Disregarding sign, the average intercorrelation among the five measures is exactly 0.10. However, since the hallmark of psychometric *g* is a *positive* correlation among all tests (Spearman, 1923, 1927), a phenomenon referred to by Thurstone (1947) as the *positive manifold*, it is more appropriate to assume that the negative correlations amount to 0.00, leaving an average positive intercorrelation of 0.07. Thus, one of the prerequisites for assuming the existence of a general or *g* factor, a positive manifold, was absent, precluding the step usually taken next in psychometric studies, namely, factor analysis.

The low level of intercorrelation in batteries of tests given to normal animals of the same species is not an original finding (e.g., Commins, McNemar, & Stone, 1932; Lashley, 1929; Livesey, 1970; McCullock, 1935; Thorndike, 1935; Warren, 1961) especially when paradigms are noncomplex and/or markedly different and subjects tend toward homogeneity (e.g., albino rats). Conversely, high intraindividual consistency will result from correlations among tasks that are slight variations of the same paradigm, as has been shown for the detour problem, confirming similar observations by Tolman and associates (e.g., Davis & Tolman, 1924), Tryon (1931), Warren (1961), and Livesey (1970). The most frequently cited example is that of mazes, especially very complicated ones, where intercorrelations in the 0.80 range have been found (e.g., Tryon, 1931). On the other hand, simple discrimination tests, mediated by a single sensory modality, tend toward zero intercorrelations with most other tests (Thorndike, 1935). We have found nothing to contradict the conclusions of these prominent figures in the history of comparative psychology, nor would we have expected to, in view of the widely varied tasks selected for study in an animal presumably bred for behavioral *homogeneity.*

At this juncture, it is well to go back to the focus of our theoretical concern, the nonspecific mechanism, and recall that this mechanism could not have been initially identified without the imposition of brain lesions, experimentally rendering the rat population *heterogenous.* Our next step was to conduct a correlational study on the data derived from a heterogeneous population.

Analysis of Brain-Damaged and Control Rats

The data from the 75 sham-operated animals were combined with those from 349 additional animals, the latter group all having sustained bilateral lesions to one of 49 of the 50 brain structures described in Chapter 3 (complete data on all tasks were not available for the lateral midbrain group). The test battery included all measures listed in Table 6.1. Additionally, since there was considerable variability in the frequency of foot shocks required to induce some lesioned animals to leave the start box in the discrimination and maze tests (controls having shown virtually no variability), a sixth measure, "number of shocks," was added to the battery. Table 6.4 gives the intercorrelation matrix for the six variables.

Even a cursory inspection of Table 6.4 reveals a vastly different degree of intraindividual consistency from that seen in Table 6.3. First, the results are *all positive.* Second, the majority of them are *statistically significant.* Finally, beyond mere statistical significance (which to some extent is the product of a large N) emerges the finding that in some cases sizable portions of the variance in one measure can be accounted for by another measure. The average intercorrelation is 0.21, itself statistically significant ($p < 0.01$), amounting to a rather marked increase over the average intercorrelation of 0.07 calculated from the matrix derived from scores of control animals only. The fact that these coefficients are derived from performance on a bat-

TABLE 6.4. Matrix of Intercorrelations for 424 Animals on Six Performance Measures[a]

Measure	1	2	3	4	5	6
1. Detours	1.00	0.49	0.44	0.27	0.11	0.12
2. Maze		1.00	0.23	0.19	0.25	0.25
3. Response latency (detours)			1.00	0.30	0.07	0.06
4. Shocks (discrimination/maze tests)				1.00	0.17	0.11
5. Inclined plane					1.00	0.16
6. Black–white discrimination						1.00

[a]For $N = 424$, with six variables, the following significance levels apply: $p < 0.05$, $r = 0.165$; $p < 0.01$, $r = 0.192$.

tery of *varied behavioral tests* makes this finding additionally remarkable.

Since all correlations were positive and generally significant, we were encouraged to proceed with factor analysis. The statistical procedure used to initially explore this data set, a key cluster analysis (Tryon & Bailey, 1970), was intentionally selected from the myriad available techniques because the underlying principle, "domain sampling," obviates reliance on many of the restrictive assumptions of conventional factor analysis (Tryon, 1959). Mathematically, key cluster analysis is extremely straightforward, based on enhanced reliability of measurement as intercorrelated variables are aggregated. Furthermore, the direct clustering solution does not force an orthogonal solution on data that may be naturally oblique. Finally, Tryon's communality exhaustion criterion for limiting the number of dimensions extracted seems more reasonable than extraction termination rules based on more arcane mathematical conventions.

The result of key cluster analysis was the extraction of two factors, together accounting for 97% of the total communality (i.e., common factor variance), with the first factor representing 91% and the second an additional 6% of the variance. The factor loading matrix is shown in Table 6.5.

Table 6.5. Matrix of Factor Loadings for Six Performance Measures on Two Factors Derived from Key Cluster Analysis

Measure	Factor I	Factor II	Communality
Detours	0.78	0.25	0.62
Maze	0.60	0.30	0.50
Response latency (detours)	0.58	0.32	0.34
Shocks (discrimination tests)	0.39	−0.29	0.23
Inclined plane	0.25	−0.24	0.19
Black–white discrimination	0.23	−0.03	0.14

From the factor loading matrix it is apparent that the first factor (I) is a general factor on which all measures have significant positive loadings. This holds true to a greater extent for the complex detour and maze problems than for the intuitively simpler sensory discrimination tasks.

On first glance, the thought might arise that Factor I is an instrument factor, in that the response latency and detour error measures are two aspects of performance on the same basic problem. Alternatively, Factor I could be presumed to reflect appetitive drive, since response latency, which has the third highest loading on the factor (0.58), very likely represents a measure of motivation for food and water. The argument here would be that solving the detour problem itself is simply a reflection of appetitive drive, since both latency and error scores load so highly on the same factor. However, both of these arguments are discredited by the fact that maze performance, measured in a *different apparatus* and run under a *different motivational condition* (i.e., shock) has the second highest loading on Factor I. Thus, Factor I appears to qualify, not only psychometrically, but logically, as a *general performance factor.*

It should also be noted that all dependent measures, when unrotated, have positive loadings on this first factor. This constitutes another criterion for Spearman's *g* in that this phenomenon could not be obtained without the presence of a common underlying characteristic, which is tapped to some extent by all tests.

The next step was to determine the stability (or reliability) of the first factor. Tryon (e.g., Tryon & Bailey, 1970) believed that one method of determining stability of the factors extracted was to be found in a "reproducibility coefficient," i.e., the ability of the factor to generate the original raw scores. In this instance, Tryon's coefficient was found to be 0.92 for Factor I, indicating a more than satisfactory degree of internal reliability. But this does not counter the most common and damning criticism leveled at factor analysis (and factor analysts!), that being the concern that factors are arbitrarily determined, hence unstable across extraction procedures. To handle this issue, we also performed *principal factor* (Frane, Jennrich, & Sampson, 1983) and *principal components* (Harmon, 1967) analyses, each procedure set to extract factors with eigenvalues equal to or greater than 1.00. The result was that the first principal factor, unrotated, accounted for 80% of the variance, while the first principal component, unrotated, accounted for 67% of the variance. As shown in Table 6.6, the factor loadings are very similar

Table 6.6. Factor Loading Matrix for Six Measures on the First Factor Identified by Three Different Extraction Methods

	Extraction method[a]		
Measure	KC	PC	PF
Detours	0.78	0.79	0.77
Maze	0.60	0.74	0.65
Response latency	0.58	0.59	0.47
Shocks	0.39	0.54	0.39
Visual discrimination	0.25	0.49	0.35
Kinesthetic discrimination	0.23	0.41	0.28

[a]KC = key cluster analysis; PC = principal components analysis; PF = principal factor analysis.

to those calculated in our original key cluster analysis. Moreover, the rank orders of the six tests, in terms of magnitude of their loadings on Factor 1, are identical.

Thus, there is ample evidence of a robust first factor that accounts for most of the variability in a set of assorted problem-solving performances. What, then, is the relationship of this factor to Spearman's *g?*

It is our contention that we have, in fact, demonstrated the existence of *g* (general intelligence), at least as it has been traditionally derived, in the problem-solving performance of the white rat. In addition to the strong psychometric evidence to support this position, other features of our findings are worth noting. Most important among them is the fact that the test with the highest loading on Factor I, irrespective of extraction procedure, is the *detour* problem. We have written earlier in this volume and elsewhere (e.g., Thompson *et al.*, 1984, 1989b) on the nature of the detour-type problem, noting that the task appears to require the participation of a number of neural problem-solving mechanisms, unique among them being *response flexibility.* Paraphrasing Porteus (1950), the problem requires the animal to "do something when it doesn't know what to do."

Tasks of the detour type tend, in phylogeny, to distinguish the problem-solving abilities of higher-order mammals from those of simpler organisms. Even among chimpanzees, Kohler (1925) noted that such "insightful" behavior differentiated the brighter from the less intelligent members of his colony. For Kohler, tests of "insight" required: (1) a sudden transition from helplessness to mastery; (2) good retention of the solution; and (3) good transfer. Put more succintly, Kohler stated that insightful behavior (which he equated with intelligence) "takes into account from the beginning the lay of the land, and proceeds to deal with it in a single, continuous and definite course" (p. 190). Wertheimer (1959), also a member of the Gestalt school, believed that intelligent behavior is characterized by productive (i.e., insightful) rather than by reproductive (i.e., memorial) thinking. More recently, considerable evidence from the newer cognitive psychology suggests that the insight view of intelligence held by the Gestalt school corresponds to the *g* notion of differential psychology (Sternberg, 1985).

It is also possible to differentiate the cognitve skills of older from younger children by using problems requiring insightful solutions (Berlyne, 1970; Maier, 1936; Piaget, 1952, 1970; Viaud, 1960). Maier (1936) felt that such problems differentiated "learning" from "reasoning" (only the latter being a function of chronological age). Jensen (1980) notes that detour problems have especially strong loadings on general intelligence in that they require the capacity "to size up the elements of a problem situation," a statement remarkably similar to Kohler's "lay of the land." And certainly no discussion of this topic would be complete without noting that the hallmark of Cattell's (1963, 1971) "fluid intelligence" (vs. "crystalized intelligence") is reflected in performance on tasks which, like the detour problems, have relatively low informational content but demand the ability to grasp relationships between relatively simple elements.

At this point, a critic might suggest that the *g* factor isolated in this particular battery of tests might not be the same as that derived from a battery of different tests. In fact, such a criticism would be well founded if we had analyzed only performance on a battery composed of *non-problem-solving* tests, for example, a battery confined to measures of speed and agility, or habituation problems. But, when the battery of tests employed involves problem-solving (a point amply treated in Chapter 1), or so-called "cognitive" tasks, we have repeatedly found detour problems to have the highest loadings on the first factor extracted. Consider Table 6.7, for example, which consists of an unrotated principal components analysis of eight tests commonly thought of as measures of response inhibition. The subjects were 43 animals, either sham-operated or lesioned bilaterally in one of the following structures: parietal cortex; globus pallidus; median raphe; dorsal hippocampus; or amygdala (see Thompson et al., 1989b).

Again, a strong first factor was extracted, accounting for 57% of the variance, and the detour task had the highest load-

Table 6.7. Unrotated Principal Components of Eight Tests of Response Inhibition[a]

Test	Factor I	II	III
Detour	0.78		
Activity	0.77	−0.42	
Visual habit reversal	0.73	0.47	
Passive avoidance	0.71		
Spatial habit reversal	0.59	0.64	
Spontaneous alternation	0.55		0.72
Go–no go	0.46	0.47	−0.54
Extinction	0.41	−0.53	−0.33

[a]Correlations of less than ± 0.30 have been omitted for ease of interpretation.

ing (0.78). A similar finding was obtained when the three detour tasks were embedded in a battery of eight problem-solving tests given to 71 animals that had sustained asymmetrical lesions to various brain structures. In this instance, first factor loadings of 0.87, 0.46, and 0.79 were found for Detours A, B, and C, respectively. In yet another experiment, tests of response inhibition, simple sensory discrimination, and mazes, as well as two types of detour problems (the climbing Detour Problem A already described and one which required burrowing in sawdust underneath a barrier), constituted the test battery. The subjects were 108 animals, 92 of which were lesioned bilaterally in 1 of 49 brain structures. First factor loadings of 0.74 were found for both types of detour problems, while loadings for other tests were smaller (e.g., maze loading = 0.58). Thus, it appears that some general problem-solving capacity of the animal, common to a wide variety of tasks, but which somehow tends to be most closely linked to performance on detour-type problems (and mazes to a lesser extent), will be found when batteries of various problem-solving tests are factor-analyzed.

To obtain a *g*, however, one additional requirement must be fulfilled, at least with respect to the rat; namely, variability in performance must be experimentally induced. As noted earlier, the *g* factor does not emerge when the performance of unlesioned animals is analyzed. The effect of 49 different brain lesions on performance variability is amply illustrated by Figure 6.1, which depicts separate frequency distributions for lesioned and unlesioned animals on the first factor. The factor scores in Figure 6.1 have been converted to standard T-scores, with a mean of 50 and standard deviation of 10. All 75 (100%) unlesioned (control) animals equal or exceed a cutoff score of 52, while only 155 (44%) of the 349 lesioned animals match or exceed this same score (i.e., would be false negative for brain damage).

Before looking at the correspondence of the general factor with individual lesion placements, the presence of other factors must be considered. Returning once again to Table 6.5, a second factor was identified. Recall that this factor accounted for only about 6% of the variance (14% using the principal components method). It is obvious that this factor is a *group* factor, i.e., accounting for a variance component generated by more than one test, rather than a *specific* factor, i.e., accounting for variance peculiar to only one test. It is otherwise rather hard to interpret. Figure 6.2 graphically depicts the unrotated loadings for the six measures in two-factor space, while Figure 6.3 shows the best approximation of Thurstone's (1947) simple structure criterion, finally achieved by a direct quartimin rotation (Frane *et al.*, 1983), following application of a variety of other rotational schemes.

Unrotated, the second factor appears to depict a group factor common to the detour, maze, and response latency measures (although loadings for these tests are clearly of a lesser magnitude than corresponding loadings on the *g* factor), and not common to the discrimination learning tasks. The unrotated factors are necessarily uncorrelated. The result of rotation (Figure 6.3) is to create a first factor on which the two motivation measures (response latency and shocks) line up with detour performance, while the second factor is clearly defined

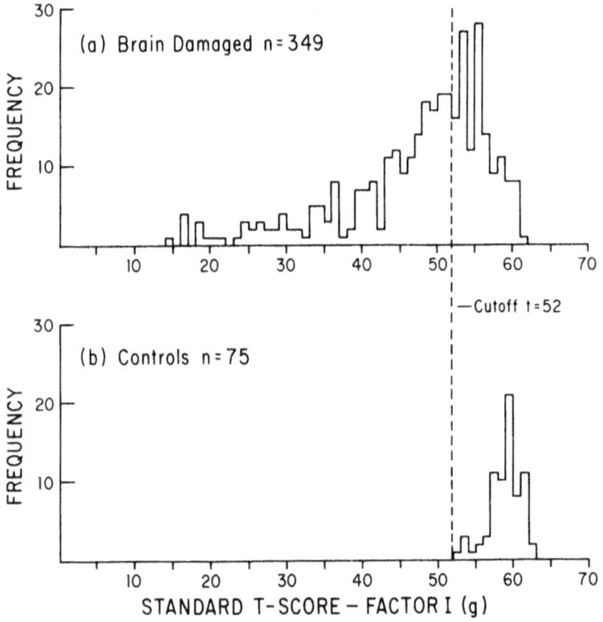

FIGURE 6.1. (a) Frequency distribution of standard T-scores (mean = 50; S.D. = 10) on Factor I for 349 rats lesioned at an age of approximately 21 days in one of 49 different brain sites. (b) Frequency distribution of T-scores for 75 sham-operated controls.

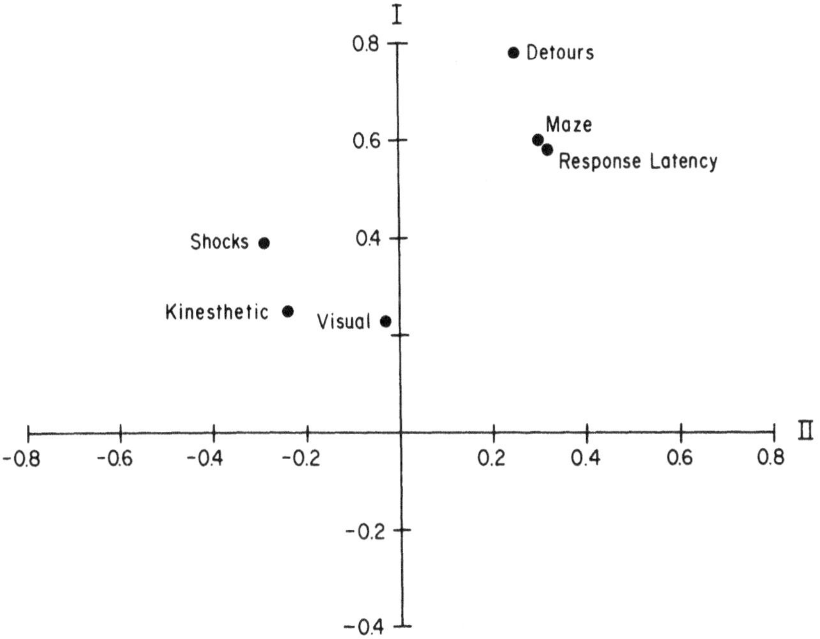

FIGURE 6.2. Unrotated positions of six dependent measures in two-factor space (Factor I = *g*; Factor II = motivation).

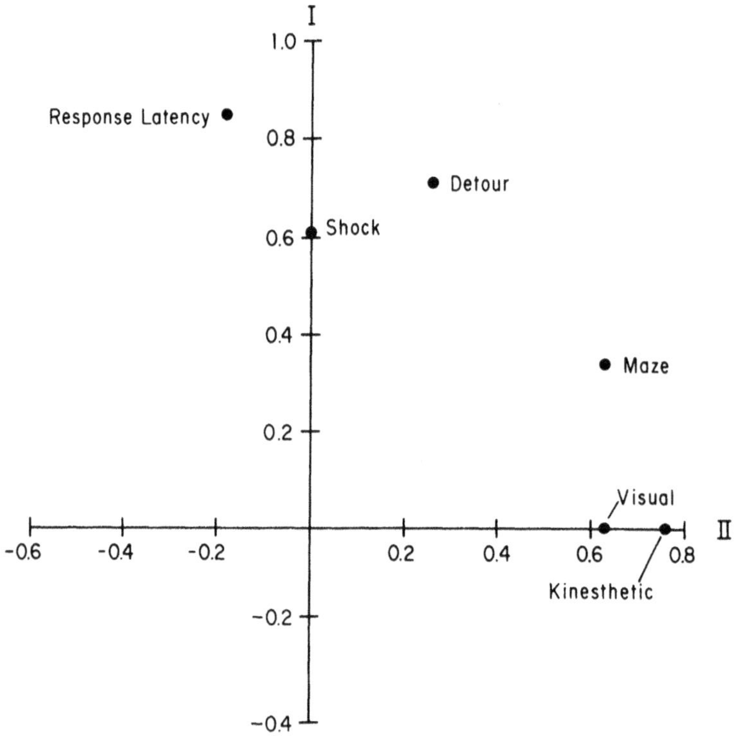

FIGURE 6.3. Positions of six dependent measures in two-factor space following rotation to best approximation of Thurstone's (1947) "simple structure." Solution is achieved by direct quartimin solution. (Compare with Figure 6.2.)

by discrimination learning, with maze performance now having coequal loadings on the two factors. At this point, the factors are modestly correlated ($r = +0.24$).

First and foremost, the search for additional factors points to the fact that other performance features tend to be "swamped" by the presence of a rather overwhelming general factor, so that variance attributable to other components becomes difficult to interpret. The rotational exercise does indicate that the response latency and shock measures, which we view as indices of motivation, should command some attention in their own right.

Factors and Lesions

Eysenck (e.g., 1961, 1982b; with Barrett, 1985) has long held that the results of factor analysis were virtually worthless without subjecting the derived factors to validation against external criteria, a viewpoint that has been influential in guiding one of us (F.M.C.) to perform factor and cluster analytic studies of humans with brain lesions (e.g., Blakeley, Crinella, & Fisher, 1985; Crinella, 1973; Crinella & Dreger, 1972). A similar view is held by Jensen (e.g., 1985) who advocates studies of the biological correlates of *g*.

Using a form of "extension analysis," in which external variables are defined in terms of factorial space determined in the main analysis (Gorsuch, 1983), we conducted yet another analysis, this time employing a "sleeper" program so that the 49 binary (i.e., present vs. absent) brain lesion variables did not influence the initial factor structure. We then activated the "sleeping" variables so that the factor loading for each lesion could be determined.

For this particular analysis, it was decided on both logical and empirical grounds to preset the factor structure. The first factor (I) was, of course, the *g* factor already identified. The second factor (II) was a "motivation" doublet, composed of scores on response latency and foot shock. The third factor (III) was a visual discrimination factor, composed entirely of performance on the black–white discrimination test, while the fourth factor (IV) was a kinesthetic discrimination factor, based on performance on the inclined plane test.

The rationale for the first factor needs no further elucidation. The second factor, motivation, is included because it was possible, with application of rotational schemes, to demonstrate an axis on which motivational, vis-à-vis cognitive, measures were virtually colinear. Furthermore, some of the lesions we have investigated clearly produced *motivational* but not *cognitive* defects, and vice versa. Additionally, as mentioned in Chapter 1, there are those who ascribe general learning deficits to motivational influences, whereas others tend to emphasize the combination of motivational and cognitive factors as necessary ingredients of intelligent behavior (Zigler & Butterfield, 1968). This (somewhat forced) factor became a way for us to partition the cognitive and motivational elements in terms of performance and its neural underpinnings.

For somewhat the same reason, the visual and kinesthetic discrimination performances of the animals were arbitrarily considered as indices of specific functions, presetting third and fourth factors to be colinear with the standard scores on the black–white and inclined plane discrimination tasks, respectively. We knew beforehand that some of the lesions (e.g., eye enucleation) investigated would have effects largely confined to these specific tests and no others. The tests were important in describing the effects of a few particular lesions, but their psychometric communalities were low. It should be kept in mind that this same fact would hold true for any of a number of tests given in the standard clinical neurological examination: They are for the most part always negative. However, in the rare instance of a positive result, such a finding is pathognomonic.

Table 6.8 is a modified factor loading matrix in which the upper section shows the individual test loadings for this preset factor analysis and the lower section the loadings for each of 49 lesions. As might be expected, almost no additional variance was accounted for by the two arbitrarily determined "specific" factors. For purposes of further discussion, these four factors are labeled I (*g*), II (Motivation), III (Visual), and IV (Kinesthetic), and are described below by either Roman numeral or performance component represented.

Inspection of Table 6.8 reveals several noteworthy findings. First, with respect to the external validity of the factor solution, it is clear that external criteria (i.e., lesions) are correlated in expected directions with certain factors. For example, occipital cortical damage as well as peripheral blinding have loadings of 0.42 and 0.47, respectively, on Factor III (Visual); red nucleus lesions have a loading of 0.46 on Factor IV (Kinesthetic); and posterolateral hypothalamic lesions have a loading of 0.62 on Factor II (Motivation).

Second, as mentioned in Chapter 5, there are many loadings that tend to emphasize nontraditional roles for certain structures. For example, lesions of the superior colliculus and olfactory bulbs seem to have a strong effect on general problem-solving behavior (Factor I).

Third, the majority of lesions have *no significant loadings* on any factor. In fact, as shown in Table 6.8, there are a remarkable number of lesions with *negative loadings* in the matrix. This, however, is an illusory finding: Such "beneficial" effects appear only because the lesions are, *on the average, somewhat less damaging than others* on this particular set of problems and derived factors. (Thus negative loadings fall well short of an endorsement for "cognitive rehabilitation" via psychosurgery.) Only 349 animals were used to study the effects of 49 different lesion placements, meaning that an average of seven animals composed each lesioned group. Moreover, given the differential survival rates for some lesions, five groups had no more than four animals (in one case three). In the computational methods employed, such overwhelmingly skewed binary (i.e., *1* or *0*) distributions would result in smaller correlation coefficients and even more miniscule communalities (the sum of squared loadings for all factors) for any given lesion in a factor analysis. In fact, the communalities (denoted in parentheses to indicate that these are external criterion, rather than performance, variables) for each lesion in Table 6.8 averaged only 0.08 per lesion, with a range of 0.03 (dorsal raphe nuclei, ventromedial hypoth-

TABLE 6.8. *Factor Loading Matrix for 349 Lesioned and 75 Control Animals (Combined) on Six Variables*[a]

Variable	Factor loading I	II	III	IV	Communality
Test					
Detours	74	63			62
Maze	74	37	48	45	61
Response latency	46	62			41
Shocks	31	52			29
Black–white discrimination			51		26
Inclined plane discrimination			31	56	31
Lesion					
Habenula			−27	−14	(07)
Ventral caudatoputamen			−15		(05)
Ventromedial thalamus	12		−10	−10	(07)
Dorsal raphe nuclei					(03)
Ventrolateral thalamus			12	19	(06)
Midbrain central gray					(04)
Posterolat. hypothalamus	30	62			(22)
Entopeduncular nucleus				−21	(08)
Rost. mid. retic. form.			−19	50	(15)
Anterior thalamus	17		−28		(08)
Medial dorsal thalamus		−10	−28		(08)
Parafascicular nucleus		10	−15	10	(06)
Lateral thalamus			−12	14	(05)
Amygdala	−11		−17	−22	(07)
Central lateral thalamus					(03)
Red nucleus		14	−13	46	(14)
Peripeduncular nucleus	−12	−13			(07)
Rostral caudatoputamen				−14	(06)
Subthalamus		36	26	−12	(17)
Dorsal caudatoputamen			22		(07)

(continued)

TABLE 6.8. (*Continued*)

Variable	Factor loading I	II	III	IV	Communality
Substantia nigra			14		(04)
Globus pallidus	−10	−12			(05)
Pontine ret. form.			16	19	(06)
Median raphe nuclei			29	19	(08)
Dorsal hippocampus	21	13			(09)
Dorsal frontal cortex		−10			(04)
Parietal cortex	24	13	36	20	(10)
Occipitotemporal cortex	20		42		(14)
Anterior cingulate cortex			−10	15	(06)
Posterior cingulate cortex	20		−19	14	(12)
Superior colliculus	24		13		(13)
Ventral pallidum			−14	−11	(06)
Ventral tegmental area	11	39		31	(18)
Nucleus accumbens			−10		(04)
Anterior pretectal nucleus					(08)
Medial pretectal area					(04)
Ventral frontal cortex				−12	(04)
Entorhinal cortex			39		(11)
Ventral hippocampus			21	11	(06)
Septofornix				12	(06)
Central tegmentum				39	(11)
Olfactory bulbs	18		15		(10)
Mamillary bodies	18	18	−12		(11)
Ventromedial hypothalamus			10		(03)
Anterior hypothalamus			18	−11	(06)
Cerebellum			−16	14	(05)
Inferior colliculus			−22	−23	(08)
Ventral thalamus			−12		(05)
Eye enucleation (blinded)			47		(13)

[a]Decimals are omitted for ease of interpretation. Loadings of $< \pm 0.30$ for performance variables and $< \pm 0.10$ for lesions have been omitted for ease of interpretation. For $N = 424$, and 196 (i.e., 49×4) factor-by-lesion correlations, the following significance levels apply: $p < 0.05$, $r = 0.291$; $p < 0.01$, $r = 0.315$.

alamus, and central lateral thalamus) to 0.22 (posterolateral hypothalamus).

Thus, the rather modest factor loadings for individual lesions must be viewed as artificially deflated, although their relative strengths serve a clear interpretive purpose. One way of viewing the communality of a lesion might be as an overall "devastation index," that is, the extent to which a particular lesion is responsible for disruption of overall performance on the test battery, at least as it is characterized by this particular four-factor hyperspace.

Table 6.9 encapsulates the major lesion by factor findings. Considering the fact that correlations, hence communalities and factor loadings, are somewhat deflated, we determined that any lesion with a communality greater than 0.10 or lesion with a factor loading greater than 0.20 was worthy of listing in Table 6.9. Even employing this sensitive standard it is obvious from the table that only 16 of 49 lesions have appreciable loadings on any of the four factors (several are listed on more than one factor); 14 lesions had communalities that exceeded 0.10, qualifying in our estimation as "devastating" lesions.

With respect to the first (*g*) factor, these findings tend to emphasize the importance of only a few structures in the highest forms of problem-solving behavior in the rodent. These structures are found at *all levels of the brain*, including the neocortex (parietal and occipitotemporal cortex), limbic forebrain (posterior cingulate cortex and dorsal hippocampus), hypothalamus (posterolateral hypothalamus), and midbrain (superior colliculus). Surprisingly, except for the superior colliculus, the structures composing the nonspecific mechanism (indicated by asterisks in Table 6.9) were not found to play a prominent role in psychometric *g*. The ventral tegmental area, another component of the nonspecific mechanism, appears prominently in Factor II, our motivational measure (which was somewhat arbitrarily teased out of Factor I). It may thus be via its motivational properties that the ventral tegmental area plays a role in the nonspecific mechanism. Two other nonspecific mechanism structures, the median raphe and dorsal caudatoputamen, appear to exert their effects via the visual system, judging

TABLE 6.9. *Major Lesion by Factor Findings*[a]

Lesion	Loading
Factor I	
1. Posterolateral hypothalamus	0.30
2. Superior colliculus*	0.24
3. Parietal cortex	0.24
4. Dorsal hippocampus	0.21
5. Occipitotemporal cortex	0.20
6. Posterior cingulate cortex	0.20
Factor II	
1. Posterolateral hypothalamus	0.62
2. Ventral tegmental area*	0.39
3. Subthalamus	0.36
Factor III	
1. Eye enucleation (blinding)	0.47
2. Occipitotemporal cortex	0.42
3. Entorhinal cortex	0.39
4. Parietal cortex	0.36
5. Median raphe nuclei*	0.29
6. Subthalamus	0.26
7. Dorsal caudatoputamen*	0.22
8. Ventral hippocampus	0.21
Factor IV	
1. Rostral reticular formation	0.50
2. Red nucleus	0.46
3. Central tegmentum	0.39
4. Ventral tegmental area*	0.31
5. Parietal cortex	0.20
"Devastation index" (communality)	
1. Posterolateral hypothalamus	0.22
2. Ventral tegmental area*	0.18
3. Subthalamus	0.17
4. Rostral reticular formation	0.15
5. Red nucleus	0.14
6. Occipitotemporal cortex	0.14

(*continued*)

TABLE 6.9. (Continued)

Lesion	Loading
7. Superior colliculus*	0.13
8. Eye enucleation	0.13
9. Posterior cingulate cortex	0.12
10. Central tegmentum	0.11
11. Mamillary bodies	0.11
12. Entorhinal cortex	0.11
13. Olfactory bulbs	0.10
14. Parietal cortex	0.10

[a]Lesion correlations with factors of $< \pm 0.20$ are not listed for brevity and ease of interpretation. Lesion communalities of <0.10 are not listed. *Indicates that lesion is also part of the nonspecific mechanism.

from their loadings on Factor III. Interestingly, the occipitotemporal cortex, which has an even more striking role than either of these two structures in both *g* (Factor I) and visual discrimination (Factor III), is not a component of the nonspecific mechanism. Finally, the remaining structures of the nonspecific mechanism–globus pallidus, substantia nigra, ventrolateral thalamus, and pontine reticular formation—do not emerge as strong external correlates of *g*. On the contrary, these four structures are characterized by rather minimal communalities.

Discussion

Does the finding of minimal overlap of neural substrates invalidate the reality of either a nonspecific problem-solving mechanism or that of a psychometric *g*? We do not believe this to be the case. Rather, psychometrically derived *g* and the nonspecific mechanism are simply not the same. What remains to be determined is the relationship of these two systems, each emanating from a different experimental perspective, to Spearman's hypothetical construct, i.e., a biological process common to all problem-solving performances.

Psychometric *g*

Spearman believed that the invariant *psychometric* finding of a factor common to all tests (*g*) constituted irrefutable evidence for a biological trait, differentially distributed among humans, "intelligence." We initially set about to determine whether something equivalent to the *g* factor found in batteries of cognitive tests administered to humans could be found in analogous performances of a population of rats (rendered heterogeneous by a variety of lesions). It was concluded that such a factor was indeed present, based on the finding of a robust first factor with the requisite psychometric properties.

We also saw that the *g* factor was sensitive to only a few lesions, thus supporting a localized "functional system" view of *g*, rather than a *g* that reflects some property of the whole brain, as suggested by Spearman, Eysenck, Jensen, and others. That is, the process (or processes) unique to those behaviors which, over 75 years of psychometric investigation, have come to be regarded as "intelligent" appears to be mediated in the rat by an ensemble of six brain structures, namely, the parietal cortex, occipitotemporal cortex, posterior cingulate cortex, dorsal hippocampus, posterolateral hypothalamus, and superior colliculus. Interestingly, four of these six structures are related to either the cerebral cortex or limbic forebrain, the two major divisions of the brain most frequently linked to higher mental functions in humans, including thinking, problem-solving, and memory. This ensemble of six structures will henceforth be termed the "psychometric *g* mechanism."

Even in the absence of whole brain involvement, which Spearman hypothesized to be the case, the essence of the Spearman *g* notion might be preserved if the brain structures critical for the psychometric *g* mechanism were also found to be important for all tasks. But this is decidedly not the case. For example, the posterolateral hypothalamus has no significant effect on the inclined plane problem; the parietal cortex does not emerge as critical in puzzle-box problem-solving; the dorsal hippocampus is not involved in discrimination (visual or inclined plane) or puzzle-box problems; and lesions of the occipitotemporal cortex fail to produce defects on the inclined plane problem or two of the three detour problems. Thus, while the structures constituting the psychometric *g* mechanism have noteworthy loadings on the *g* factor, they are clearly not critical for the acquisition of many problem-solving tasks.

Viewing the six brain structures that participate in the psychometric *g* mechanism in the context of the neural problem-solving mechanisms described in Chapter 5, one fact becomes abundantly clear: *Almost every brain structure implicated in psychometric* g *participates in a different neural mechanism.* The visuospatial attentional mechanism is represented by the parietal cortex; the visual discrimination mechanism is represented by the occipitotemporal cortex and posterolateral hypothalamus; the vestibular–proprioceptive–kinesthetic discrimination mechanism is represented by the posterior cingulate cortex; the place-learning mechanism is represented by the dorsal hippocampus; and the nonspecific mechanism is represented by the superior colliculus. And, as shown earlier, the detour and maze tests, with by far the highest loadings on psychometric *g*, cannot be learned without the participation of the foregoing five mechanisms.

These findings are precisely what one would predict from the neoassociationist view of psychometric *g* advanced several years ago by Godfrey Thomson (1951) and recently reintroduced in the form of an elegant systems theory thesis by Detterman (1987). For them, *g* results from the fact that each mental test requires the recruitment of a number of distinct abilities (bonds) and that some of these abilities are common to most tests. Like Guilford (1964), Detterman believes that if tests are selected so that each samples only a single element of the subject's cognitive domain, essentially zero intercorrelations would be found and *g* would not emerge. On the other hand, high intercorrelations and a robust *g* would be the outcome if the chosen tests tap multiple elements of the subject's cognitive domain. Furthermore, tests that sample many elements will have higher *g* loadings than those that sample fewer elements. Returning to our rat data, it would follow from the foregoing conceptualization that a significant *g* would appear in the factor analysis since our tests sampled a number of common abilities (such as visuospatial attentional, visual discrimination, and place learning skills) whose distinct neural substrates were identified in Chapter 5. Furthermore, it would follow that the detour and maze tasks would have higher *g* loadings than the discrimination tasks because the former tap more (from five to six) abilities than the latter (only three). Figure 6.4 is a graphic representation of this particular view of *g*. It should be clear that tests with higher *g* loadings (indicated in parentheses) sample more of the basic neural mechanisms (represented by circles) than tests that sample fewer neural mechanisms.

If the simultaneous sampling of at least five neural problem-solving mechanisms is responsible for the high *g* loadings of the detour and maze tests, one might then wonder why *all of the structures* constituting each mechanism do not have high *g* loadings. For example, the amygdala and habenula are involved in the place-learning mechanism, but have negligible *g* loadings. Conceivably, the six brain structures that have high *g* loadings may simply be more critical than other structures within the same mechanism (i.e., more central to that mechanism) for the performance of detour and maze tasks.

Nonspecific Mechanism or Biological *g*

Returning to the relationship of the nonspecific mechanism to Spearman's *g*, several features of this system need reemphasis: (1) a lesion to any one of eight structures composing this mechanism will significantly impair performance on all problems; (2) a localized functional system, rather than a "whole brain" process, is implicated, and (3) there appears to be no known or describable contribution (cognitive component) of this mechanism to problem-solving (hence its designation as "nonspecific"). Thus, whether three, four, or six specific mechanisms are involved in mediating a particular test performance, this real, but indescribable, mechanism will apparently be included within the aggregation. The simple logic of this position is graphically illustrated in Figure 6.5 by a Venn diagram, with each circle representing a particular test. In the

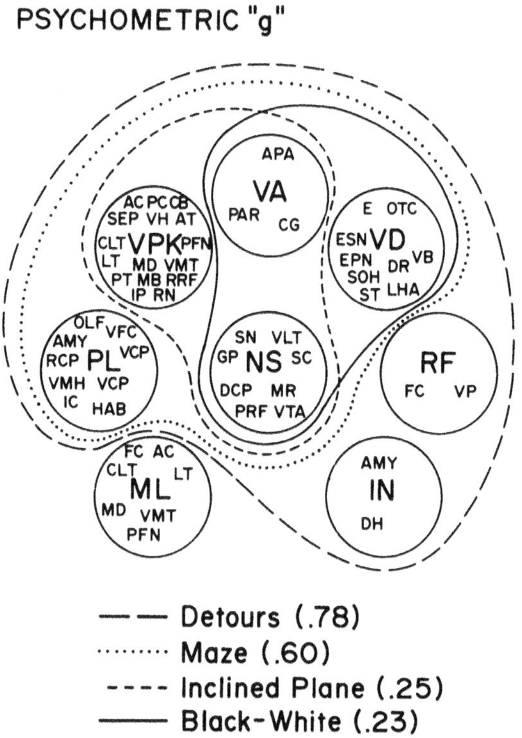

FIGURE 6.4. Domain sampling characteristics of four tests (represented by various broken lines) that lead to differential loadings on the *g* factor. IN = inhibition (from Thompson *et al.*, 1989b); ML = motor learning mechanism; NS = nonspecific mechanism; PL = place-learning mechanism; RF = response flexibility mechanism; VA = visuospatial attentional mechanism; VD = visual discrimination mechanism; VPK = vestibular–proprioceptive–kinesthetic discrimination mechanism.

segment of the diagram common to all tests, the eight structures comprising the nonspecific mechanism, and no others, are to be found. Thus, on a logicodeductive, as opposed to a psychometric, basis a mechanism common to all problem-solving performances can be identified. Furthermore, it can only be recognized when the performance contributions attributable to various brain regions are analyzed in this fashion, where lack of significant contribution to performance on *any test* results in the exclusion of the structure from the system. The residual of this serial subtractive process is a mechanism that is indisputably both general and biological. Given these features, the nonspecific mechanism can now be termed "biological *g*."

Psychometric vs. Biological Considerations

Recall that this particular treatment of our data was undertaken in the belief that the *g* factor, which Spearman and others had identified psychometrically, was more or less equivalent to *intelligence,* and further, that intelligence, so comprised, would ultimately have an identifiable neurobiological basis. A plausible extension of this line of reasoning was the hypothesis that the neural underpinnings of intelligence (psychometrically derived) would be found in the nonspecific mechanism, which had already been demonstrated to be involved in every problem-solving performance.

Contrary to our supposition, two distinct processes were identified, either of which could justifiably lay claim to the label "intelligence" in view of the fact that the first fits the traditional psychometric criteria for the construct while the second fulfills a biological or adaptive definition. Each process is subserved by a different set of brain structures, which have now been termed the psychometric *g* mechanism and the biological *g* mechanism. But, does one of these two mechanisms, more than the other, deserve to be thought of as the neural basis of "intelligence?"

Halstead (1947, 1951) grappled with much the same question in interpreting human performance. Following factor analysis of the performance of humans with brain lesions, he sug-

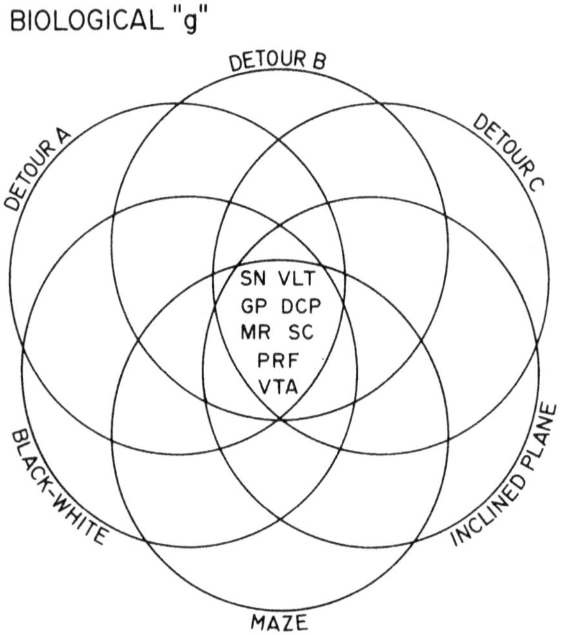

FIGURE 6.5. Venn diagram illustrating the logicodeductive method by which the nonspecific mechanism, now termed "biological *g*," has been identified. Each circle represents a performance test and lower-case letters correspond to abbreviations for brain structures given in Chapter 3. (The structures depicted in the area common to all six tests represent biological *g*.)

gested a four-factor theory of biological intelligence, with each factor tied to a region of the cerebral cortex. Such a finding led Halstead to conclude that each factor was "biologically important for the survival of the organism." Halstead went on to say:

> This biological intelligence seems to bear directly upon the capacity of man for controlled adaptability, while, over a considerable range, psychometric intelligence, as measured by the intelligence quotient, does not. It is possible that suitable methods can be devised for measuring biological intelligence in infrahuman animals, although the necessary investigations largely remain to be carried out. (1951, p. 254)

Halstead was impressed by Hebb's (1945) report that patients with the major part of both frontal lobes removed had not shown significant IQ losses; yet other forms of adaptive behavior, especially those requiring rapid shifting of abstract concepts (Halstead's "A factor"), were quite defective in these same patients. As it turned out, further investigation by Halstead led him to conclude that his A factor was probably the same as Spearman's *g*, which Spearman himself had come to view as an abstraction factor.

However convincing Halstead's argument may be, the findings reported here, for the white rat, do not implicate the frontal lobe for either the psychometric or the biological *g* mechanism. This would also be true for Halstead's other factors, which were also localized in the neocortex. Furthermore, Halstead ascribed specific behavioral features to each of his factors. Thus, while Halstead differentiated between biological and psychometric intelligence, his "biological intelligence" differs from the biological *g* we have identified by virtue of both the brain structures implicated as well as the specificity of the behavioral processes described. Parenthetically, this gap between Halstead's views and our results is at least partially bridgeable if it is assumed that in the course of evolution from rodents to humans the nonspecific mechanism expanded in a cortical direction to include the territory of the frontal lobes. This assumption is not altogether unreasonable when it is considered not only that the basal ganglia (which constitute the core of the rodent's nonspecific mechanism) have extensive connections with the frontal lobes, but that one of the more common functions ascribed to the basal ganglia embraces Halstead's A factor: namely, the shifting of cognitive sets (Buchwald *et al.*, 1975; Cools, Van Den Berken, Horstink, *et al.*, 1984). In any event, we must give Halstead credit for emphasizing a biologically based capacity for adaptation, which stood apart from those abilities measured by conventional IQ tests. What Halstead demonstrated was that preserved IQ did not necessarily equate to preserved survival capacity. The same line of reasoning can be applied to the white rat in the sense that a *g*-deficient animal may be impaired in the rat equivalent of an IQ test, but its survival would be more threatened by injury to the nonspecific mechanism.

Historically, of course, IQ tests were constructed to maximize psychometric *g* content, since this had proven to be the best available predictor of future academic performance (Binet & Simon, 1916). For purposes of predicting academic success, tests with high *g* loadings, such as analogies, are as yet unsurpassed (Jensen, 1980). However, in populations such as the mentally retarded, in which the prediction of survival skills tends to override concerns over future academic achievement, measures of "adaptive behavior" have been found to be much more relevant (Leland, Shellhaas, Nihara, & Foster, 1967). Adaptive behavior can be defined as "the manner in which the individual copes with the natural and social demands of his environment" (AAMD, 1961). Measures such as the American Association on Mental Deficiency (AAMD)-sponsored Adaptive Behavior Scale are structured to assess a broad range of survival skills, which have been shown to be inadequately sampled by conventional IQ tests, even though correlations in the 0.50–0.60 range are generally found between the two measures (Heber, 1962). Since correlations of this magnitude meant that adaptive behavior could be seriously miscalculated from IQ scores alone, AAMD adopted a two-dimensional rating (i.e., IQ and adaptive behavior) to establish the degree of mental retardation.

Thus, there is evidence from two distinct lines of inquiry, the first being Halstead's factorial studies of biological intelligence and the second being the AAMD adaptive behavior project, which tends to discount psychometric *g* as an index of intelligence, at least in an adaptive or survival sense.

Yet another hurdle for those seeking a correspondence between psychometric *g* and a biological substrate lies in the fact that five of the six brain structures with the highest loadings on the psychometric *g* mechanism were found to be participants in *identifiable elementary processes*. Hence, they mediate nothing "special," in a cognitive sense. Similarly, only one of the structures comprising the psychometric *g* mechanism, when lesioned, affected performance on all tests (superior colliculus). Thus, the preponderance of structures in the psychometric *g* mechanism are not critical for *all problem-solving performances*.

If nothing else, our findings, added to the aforementioned investigations, pose additional challenges for those who subscribe to the position that the *g* derived from factorization of psychometric tests will ultimately converge on a pure biological index of intelligence. In contradistinction to the psychometric findings, a logicodeductive analysis of performance deficits and lesion placements resulted in the identification of a mechanism common to *all problem-solving performances, yet exclusive of any identifiable elementary behavioral process*.

Perhaps, then, Spearman was correct in positing a special trait underlying all cognitive performance but incorrect in assuming that the behavioral manifestations of such a trait could be isolated via factor analysis. Our experience tends to confirm the views of Thomson (1951) and Detterman (1987) in that it represents yet another example where the psychometric *g* factor, once identified, could be dissected out to reveal more elementary components (in this case by virtue of the various neural mechanisms represented). Spearman, in the company of Lashley and others, would also seem to have been incorrect in assuming that *g* was a function of the activities of the cerebral cortex. The nonspecific mechanism, arguably biological *g*, rather than being directly related to the action of the neocortex, seems to be a function of the collective activities of eight circumscribed subcortical structures.

7

Interpreting the Results of the Problem-Solving Atlas

Penfield's Centrencephalic Theory

A Recapitulation

This lesion odyssey through the rat brain in search of structures critical for problem-solving activities has yielded an extraordinary amount of data, much of which is new. In view of the large number of brain sites examined and the diversity of laboratory tasks investigated, these data have the potential to offer perspectives for increasing insight into the neural mechanisms of learning. In fact, it is our view that embedded within this pool of observations dwells some of the missing links that could bridge the gap between the neurology of learning, on the one hand, and the functional organization of the brain, on the other. But first, it is necessary to review some of the conclusions bearing upon brain–learning relationships that have been drawn from the data shown in Chapter 4. Any comprehensive theory of brain function must necessarily come to grips with these conclusions.

Some Conclusions

Structures important for the acquisition of any given problem-solving task are found in widespread regions of the brain.

This conclusion, which is based on a lesion analysis of discrimination, maze, and detour problems, is straightforward and few would be expected to dispute it. Interestingly, a similar conclusion was made over two decades ago by E. Roy John (1967), after reviewing electrophysiological studies of brain activity during the establishment of learned responses, and was later reinforced by electrophysiological findings dealing with the identification of "learning units" within the brain; they were found to reside not only within the cerebral cortex, but within the hippocampus, basal ganglia, thalamus, hypothalamus, and brainstem reticular formation as well (Disterhoft & Olds, 1972; Kornblith & Olds, 1973; Linesman & Olds, 1973; Olds, Disterhoft, Segal, *et al.*, 1972).

Different constellations of brain structures are important for the acquisition of different classes of problem-solving tasks.

Comparing those regions of the brain critical for the normal acquisition of a visual discrimination problem with those critical for the normal acquisition of a maze, inclined plane, or detour problem (see Chapter 5) clearly reveals that different kinds of laboratory tasks are mediated by different ensembles of neural structures. This conclusion is also unlikely to be opposed since it is supported by the classic studies of Lashley (1929), which revealed that the brain mechanisms subserving maze learning are quite different from those subserving visual discrimination learning, as well as by recent neurobehavioral research aimed at identifying dissociable brain systems underlying different classes of learning and memory (see Squire & Butters, 1984).

There exists a group of brain structures essential for normal acquisition of virtually all types of problem-solving activities.

This conclusion rests on the findings that our rats with lesions in the vicinity of either the dorsal caudatoputamen, globus pallidus, ventrolateral thalamus, substantia nigra, ventral tegmental area, superior colliculus, median raphe, or pontine reticular formation were deficient in learning all laboratory tasks composing the test battery (discrimination, maze, detour, and puzzle-box problems). This conclusion is not in keeping with current neuropsychological views in three main respects. First, as far as can be determined, no other laboratory has reported systematic evidence favoring the existence of a group of brain structures having nonspecific functions in learning. Second, most of the regions of the brain that we found to play a general role in problem-solving have traditionally been associated with motor functions rather than with learning or memory. Finally, a nonspecific mechanism related to problem-solving is inimical to the more traditional view of learning and memory which envisages the brain as a consortium of independent systems, each of which is concerned with a special category of problem solving-activities (see, for example, Oakley, 1983b; Pribram, 1984; Weiskrantz, 1987).

Those structures of the brain important for normal acquisition of any given problem-solving task can roughly be grouped into at least two sets of neural mechanisms, one having nonspecific functions and the other having specific functions related to the cognitive, sensory–perceptual, motivational, and motor demands of the task.

This conclusion is grounded on the outcomes of the comparisons made among the lesion-defined learning systems associated with each problem-solving task (see Chapter 5). It was found that the learning system associated with the visual discrimination problem could be fractionated into a nonspecific mechanism, a visuospatial attentional mechanism, and a visual discrimination mechanism; that associated with the inclined plane problem could be fractionated into a nonspecific mechanism, a visuospatial attentional mechanism, and a vestibular–proprioceptive–kinesthetic discrimination mechanism; that associated with the maze problem could be fractionated into a nonspecific mechanism, a visuospatial attentional mechanism, a visual discrimination mechanism, a vestibular–proprioceptive–kinesthetic discrimination mechanism, and a place-learning mechanism; and that associated with the detour problems could be fractionated into a nonspecific mechanism, a visuospatial attentional mechanism, a place-learning mechanism, possibly a visual discrimination as well as a vestibular–proprioceptive–kinesthetic discrimination mechanism, and a response flexibility mechanism. This conclusion, which encapsulates the fundamental findings of the current study, is not in accord with the mainstream of neuropsychological thinking. On the other hand, by assuming that the nonspecific mechanism (which is entirely subcortical in origin) is responsible for the selective activation and orchestration of the specific mechanisms used to solve a given problem, there emerges a conceptualization of brain function virtually identical to that contained in Wilder Penfield's (1954b, 1958) theory about the "centrencephalic integrating system." This challenging theory, which proposes that the highest level of cognitive integration occurs at brainstem levels, did manage to stir some interest with the advent of research on the ascending reticular activating system (Delafresnaye, 1954), but has since been on the wane. The devastating critique of Penfield's theory by Walshe (1957) undoubtedly promoted an enduring disaffection with any kind of theory positing a brainstem integrating mechanism.

In our view, however, the overall findings of the current study warrant a reevaluation of the issue of whether or not a subcortical integrating mechanism like that envisaged by Penfield exists within the mammalian brain. While a few writers have attempted to revive a centrencephalic-type theory to explain certain functional relationships between brain and behavior (Diamond, 1980; Kilmer, McCulloch, & Blum, 1968; Klopf, 1982; Knoll, 1969; MacKay, 1966; Reader, 1974; Thompson, 1965), their efforts have largely gone unnoticed. Perhaps the main reason for this lies in the vagueness and imprecision in specifying the brain sites that make up the centrencephalic integrating system. Since Penfield's theory is basically anatomical in nature and focuses largely on the functional organization of the brain, its explanatory power will remain limited and suspect so long as the elements of the centrencephalic integrating system are shrouded in mystery. Our discovery of a nonspecific mechanism underlying problem-solving activities opens the way to at least a provisional identification of the composition of this system. As will be seen later, specifying the neuroanatomical makeup of this system renders Penfield's theory more defensible, more testable, and more applicable to findings bearing on experimental and clinical issues associated with the functional organization of the brain (see Chapter 8).

Penfield's Centrencephalic Theory

Penfield's view of brain function arose largely from clinical observations (Penfield, 1952, 1954a,b, 1958; Penfield & Jasper, 1954). He was particularly impressed by the findings that small interferences within the brainstem by either undue pressure, injury, or local epileptic discharges abolished consciousness and purposeful action, while extensive cortical excisions left these "psychic" activities intact. Another source of evidence that impelled him to favor the existence of a brainstem integrating mechanism related to the manner in which the precentral motor gyrus is activated. For Penfield, transcortical pathways did not participate in eliciting responses from the motor cortex since cortical ablations surrounding the precentral gyrus failed to compromise voluntary movement. The only alternative was that the motor cortex was activated by a ganglionic mass residing beneath the cortical gray mantle.

In light of these observations, Penfield proposed the existence within the upper brainstem (portions of the diencephalon, mesencephalon, and metencephalon) of an ensemble of nuclei and pathways that corrdinates and integrates the activities of the cerebral hemispheres. This centrally located integrating ensemble was termed the "centrencephalic system." It is important to emphasize that this system comprises two major divisions, an "essential portion" or "core" (consisting of certain brainstem regions) and the interconnecting pathways linking this essential portion with the remainder of the central nervous system (particularly the cerebral hemispheres). Functionally, this system represents

> a ganglionic area in which that stream of nervous impulses must arise that produces voluntary activity, an area in which the senso-

> ry pathways culminate in neurone circuits in which the information relative to past experience is made available, an area in which those nervous mechanisms are to be found which are prerequisite to the existence of intellectual activity and prerequisite to the initiation of the patterned stream of efferent impulses that produce the planned action of the conscious man. (Penfield, 1954b, p. 286)

This theory was adopted both as a working hypothesis with specific applications to clinical neurological cases and as a "protest" against the popular view that higher mental functions are subserved exclusively by the multiplicity of pathways within the cerebral cortex. With the fulfillment of these goals, Penfield apparently felt no need to refine his theory either by specifying the composition of the centrencephalic system or by formally addressing it to issues related to problem-solving and other cognitive activities. For purposes of exposition, however, the theory can be characterized in a visual discrimination problem-solving setting in the following way: Visual impulses are projected via the optic pathways to the occipital cortex and then are relayed to the essential portion of the centrencephalic system (hereafter referred to as the "centrencephalic core, or CC) within the upper brainstem to be incorporated with other sensory impulses relayed in a similar fashion. Because of this system's vast interconnectivity with other parts of the brain, engrams associated with this visual input are excited and different cortical and subcortical areas are called into service for further information processing. Upon integrating this processed information with information related to the motivational needs at the moment, the CC would generate and discharge a particular pattern of "volitional" impulses to the appropriate motor mechanism leading to the production of a planned response.

Obviously, Penfield's theory is no more than a bare outline of the sequence of a few relevant events that intervenes in the elaboration of problem-solving behaviors. Little consideration is given to motivational, emotional, or attentional factors that must operate in such behaviors. Moreover, as Walshe (1957) pointed out, Penfield's insistence that the various sensory cortical receiving areas are "way-stations" or "stop-overs" hardly provides any insight into the important functions that must be transacted by the cerebral cortex. Nevertheless, Penfield's theory has some merit in presenting the morphological superstructure for a conceptualization of brain function that runs the gamut from the reception of stimuli and the arousal of memory traces to the integration of cortical activities and the production of voluntary responses. This contrasts with most other neuropsychological theories, which rarely discuss cortical and subcortical structures within the broader context of brain organization.

Specifying the Centrencephalic Core (CC_r)

It should be clear by now that our lesion data on laboratory rats are readily explicable in terms of Penfield's theory. First, the nonspecific mechanism that we uncovered could constitute the neural substrate of Penfield's CC since it is subcortical in origin. Second, this nonspecific mechanism could be responsible for coactivating the various specific mechanisms (see Chapter 5) needed to solve the particular problems making up the test battery. Finally, this nonspecific mechanism, which contains elements belonging to the extrapyramidal motor system (and more specifically to the basal ganglia), could conceivably carry out the executive (motor) functions ascribed to the centrencephalic integrating system.

In light of the foregoing considerations, it is proposed that those brain sites making up the nonspecific mechanism underlying problem-solving in the rat define those sites that constitute Penfield's CC. This proposal would appear to be congruent with the spirit of Penfield's theory when it is considered that the centrencephalic system was originally envisioned as coordinating and integrating the activities of that division of the brain (cerebral cortex) most intimately linked to higher mental functions. Moreover, the lesion approach that we adopted to identify sites within the rat brain that play a nonspecific role in learning yielded an ensemble of structures (striatum, globus pallidus, ventrolateral thalamus, substantia nigra, ventral tegmental area, superior colliculus, median raphe, and pontine reticular formation) that inhabits the upper brainstem, a region of the neuraxis long thought by Penfield to contain the CC.

Granting that the dorsal caudatoputamen, globus pallidus, ventrolateral thalamus, substantia nigra, ventral tegmental area, superior colliculus, median raphe, and pontine reticular formation compose the CC (hereafter given the designation of CC_r to indicate its derivation from research on the rat), one of the more important questions to arise concerns the existence of functional interconnecting pathways between the CC_r and the neocortex which would allow the former to interact with the latter. Recent findings pertaining to the connections between the cerebral cortex and brainstem in the rat reveal a number of possible routes by which these interactions can be consummated (see Table 7.1). Furthermore, these to-and-fro connections may not be the only functional links between the CC_r and the cerebral cortex. Most of the elements of the CC_r have direct connections with one or more nuclei of the thalamus (Faull & Mehler, 1985). This poses the possibility that coordination and integration of neocortical activities by the CC_r may also be accomplished through the parallel pathways afforded by thalamocortical and corticothalamic circuitry. Penfield, of course, stressed the integrative role of the thalamus throughout his writings. However, the scheme outlined here views the thalamus not so much as a superordinate entity, but as an intermediary in the chain of events that functionally unites the cerebral cortex (and other parts of the brain) with the CC_r.

Another question of considerable significance arising from the proposed composition of Penfield's brainstem integrating area is the extent to which this particular ensemble has the potential to carry out the mosaic of functions ascribed to it by Penfield. Obviously, the CC_r cannot be rigorously evaluated in terms of its capacity to initiate (or participate in) any particular cognitive operation (e.g., perception, memory, orientation, volition, attention, learning, abstract reasoning, language), but it can reasonably be shown to contain features that are essential for the execution of two rather complex functions attributed to

TABLE 7.1. Reciprocal Connections between the Cerebral Cortex and Elements of the CC_r[a]

CC_r element	Efferent connections to	Afferent connections from
DCP		Posterior cortex: Faull *et al.*, 1986
GP	Anterior cortex: Van der Kooy & Kolb, 1985	
VLT	Anterior and posterior cortex: Herkenham, 1986	Anterior cortex: Molinari *et al.*, 1985
SN	Anterior cortex: Fallon & Laughlin, 1985	Anterior cortex: Gerfen *et al.*, 1982
VTA	Anterior and posterior cortex: Oades & Halliday, 1987; Takada & Hattori, 1987	Anterior cortex: Oades & Halliday, 1987
SC		Anterior and posterior cortex: Matute & Streit, 1985
MR	Anterior and posterior cortex: O'Hearn & Molliver, 1984	
PRF	Medial cortex: Jones & Yang, 1985	Anterior cortex: Shammah-Lagnado *et al.*, 1987

[a]For abbreviations of CC_r elements, see Chapter 3.

the centrencephalic system. First, as already pointed out, the anatomical circuits available to the CC_r could be used to program the various activities of the cerebral cortex in accordance with Penfield's theory. Second, since the CC_r consists mainly of elements of the basal ganglia and anatomically associated nuclei known to play a role in motor function (Hassler, 1977; Kitai, 1981; McGeer *et al.*, 1987), the combined activities of this neural network could conceivably serve to generate and discharge that stream of efferent impulses which Penfield claimed to be responsible for voluntary action.

Collateral support for the possibility that the CC_r could constitute the major part of a neural substrate mediating complex behavioral and cognitive functions would come from the demonstration of a correspondence between predicted disorders arising from partial lesions to Penfield's higher brainstem integrating mechanism and observed abnormalities accompanying selective lesions to elements of the CC_r in humans. A list of the more obvious predicted clinical disorders might include disturbances in the "psychic" (vigilance, arousal, or attention), motor, memory, problem-solving (learning), visuospatial, language, and possibly emotional–motivational domains. Considering only those clinical cases with relatively localized brain damage to the CC_r, Table 7.2 shows that the observed lesion-induced behavioral and cognitive disorders do indeed correspond with some of the predicted outcomes of focal injuries to Penfield's brainstem integrating system. Parenthetically, the clinicopathological findings related to "subcortical" dementia also provide evidence that the basal ganglia and associated brainstem structures may represent the principal parts of the morphological substrate of Penfield's centrencephalic system (see Chapter 8).

Finally, it must be more than coincidence that the basal ganglia, which form the major part of the CC_r, have been proposed by some theorists to serve certain cognitive functions not remarkably different from those assigned to the centrencephalic system by Penfield. For example, the basal ganglia have recently been conceived to play a role in regulating "cognitive sets" that determine whether the subject will respond

TABLE 7.2. Behavioral Deficits in Humans with Lesions to the CC_r[a]

Lesion site	Deficit	Reference
CP	Motor	Jung & Hassler, 1960
	Visuospatial	Potegal, 1972
	Language	Wallesch *et al.*, 1983
GP	Psychic	Ali-Cherif *et al.*, 1984
	Motor	Jung & Hassler, 1960
	Memory	Ali-Cherif *et al.*, 1984
	Motivation–emotion	LaPlane *et al.*, 1984
	Dementia	Perry *et al.*, 1985
VLT	Language	Gorelick *et al.*, 1984
SN	Motor	Oyanagi *et al.*, 1986
	Learning–memory	Oyanagi *et al.*, 1986
VTA	Mentation	Torack & Morris, 1988
	Memory	Goldberg *et al.*, 1981
	Motivation	Trimble & Cummings, 1981
SC	Psychic	Denny-Brown, 1962

[a]For abbreviations of lesion sites, see Chapter 3.

one way or another in the presence of a given stimulus context (Buchwald *et al.*, 1975; Cools *et al.*, 1984), to serve as an intermediary between sensory input and motor output (Lidsky, Manetto, & Schneider, 1985), to participate in selective attention (Barker, 1988) and in the retrieval of stored information (Hassler, 1980), to be necessary for the acquisition and retention of procedural knowledge (Phillips & Carr, 1987; Saint-Cyr, Taylor, & Lang, 1988), and to aid in the planning and modulation of behavioral sequences in the absence of external guidance (Stern & Mayeux, 1986).

8

Implications of Centrencephalic Theory

Traditionally, cognitive functions have been viewed as being processed almost exclusively by the cerebral cortex. In Penfield's view of brain function, however, the activities of certain brainstem formations (which we have termed the "centrencephalic core," or CC) play a dominant role in the awareness of sensory experience, in the activation of stored information, in the orchestration of the various operations of the cerebral hemispheres, in the performance of problem-solving and other intellectual skills, and in the production of the stream of efferent impulses responsible for planned action. In earlier writings, Penfield (1952, 1954a,b, 1958) conjectured that the CC comprised certain nuclei within the diencephalon, midbrain, and pons. It was later intimated that the basal ganglia may also be included within the CC (Penfield, 1975a). (Although Penfield, 1975b, subsequently divided the CC into three functional parts—the "highest brain-mechanism," the "automatic sensorimotor mechanism," and the "record of experience," we will only deal with the CC as a whole.) In order to add some degree of precision to this otherwise unstructured, but potentially testable theory, we have proposed that our lesion-identified nonspecific mechanism underlying problem-solving in the rat defines the composition of Penfield's CC. In other words, the brainstem integrating area to which Penfield referred is estimated to consist of the dorsal caudatoputamen, globus pallidus, ventrolateral thalamus, substantia nigra, ventral tegmental area, superior colliculus, median raphe, and pontine reticular formation. It should be apparent, however, that the overall composition of this integrating ensemble (now termed the CC_r) within the human (or infrahuman primate) brain may not be exactly the same as that within the rat brain. (As indicated later in this chapter, the prefrontal regions of the human brain may constitute a cortical extension of the CC_r.)

Like Penfield, we have not attempted to elaborate on centrencephalic theory. Admittedly, this theory faces serious conceptual difficulties and practically defies refinement because it essentially posits a little brain within a big brain (Walshe, 1957). At the same time, however, the fundamental idea of centrencephalic integration cannot be entirely dismissed as an alternative view of cerebral organization since there is much to recommend it. As noted by Jasper (1977), "The concept of the centrencephalic integrating system has been, and continues to be, one of Wilder Penfield's most stimulating legacies to neurology and to the neurological sciences:" (p. 1371).

Modularity, Central Systems, and the CC_r

Current neuropsychological thinking is dominated by the principle of "modularity" (Fodor, 1983; Marshall, 1984; Schwartz & Schwartz, 1984) in the sense that brain processes are conceptualized not only as being divisible into component functional parts, but as being localizable to the activities of particular regions of the brain. The contents of recent books in neuropsychology are replete with animal lesion studies and reports of brain-damaged humans demonstrating the modular organization of the brain. For example, depending on the area involved, injuries to a given part of the human brain (usually the cerebral cortex) may produce deficits in either language, visuospatial abilities, calculations, planning, short-term memory, or skilled movements. However, according to Fodor (1983), this conceptualization of brain function is incomplete. What is missing is an explanation of how the regionally localized component brain processes combine to produce such global cognitive achievements as thinking, problem-solving, reasoning, and decision-making. As far as Fodor is concerned, intelligent actions can only emerge from the operation of "central systems" which assemble the products of the component brain processes or "input systems," evaluate them in terms of information in memory, and arrive at the "best hypothesis" about the significance of the current situation. Fodor notes that while there is a neuropsychology of input systems (e.g., language, face recognition, cognitive maps, detection of melodies), there is virtually no neuropsychology of central systems, at least to the extent that thinking and problem-solving have not been modularized and that focal neocortical lesions have not been found to produce a generalized intellectual impairment. This deficiency is especially reflected in the dearth of neuropsychological studies and theories dealing with clinical conditions known to have compromised central systems, such as mental retardation (Thompson *et al.*, 1986).

Most investigators have assumed or at least have taken for granted that the physical basis of central systems (central circuits) as well as that of input systems (input circuits) are to be found almost exclusively within the cerebral cortex. According to one variant of this view (Luria, 1970), the input circuits would mainly occupy the sensory cortical projection areas (primary zones) and a subdivision of the sensory cortical association areas (secondary zones), while the central circuits would largely reside within the tertiary zones of the sensory cortical association areas and the prefrontal cortex. Another variant of this view, which is based on the findings that the primary sensory cortical projection areas serve both specific and nonspecific functions (Lashley, 1943; Lubar *et al.*, 1967; Orbach, 1959; Thompson, 1982b; Winer & Lubar, 1976), would likewise envisage the input circuits as being localized to certain cortical areas, but would instead picture the central circuits as being diffusely distributed throughout the cerebral cortex (Lashley, 1943). Both conceptualizations presume that the multiplicity of pathways within the cerebral cortex is sufficient to mediate those complex cognitive functions that Fodor has ascribed to the central systems.

Evidence favoring this neocortical model of central systems is practically nonexistent. In fact, both past and present data on animals with lesions to certain parts of the brain can be interpreted as being inimical to this model and to any other model proposing that the plethora of corticocortical connections is sufficient to support central processes involved in problem-solving, decision-making, and the like. Lashley (1950), for example, showed that knife cuts through the cortex and underlying fibers (the incisions averaged half of the entire length of the cerebral hemispheres) failed to impair acquisition of a complex maze in rats (see also Sperry, 1961). More recently, Oakley (1981) has reported that rats can solve a two-choice visual discrimination problem and its subsequent reversal despite being deprived of their cerebral cortex (see also Oakley, 1983a). In a series of experiments performed at Louisiana State University, one of us (R.T.) found that decorticate rats could be successfully trained (or retrained) on such problem-solving tasks as lever pressing for a sucrose solution in a Skinner box and displacing a stimulus card or rotating a butterfly latch to obtain water. Even instances of preserved associative learning capabilities in anencephalic, hydranencephalic, and microcephalic humans have been documented (Goldstein & Oakley, 1985).

Although the foregoing data conflict with the neocortical model of central systems, they are congruent with the alternative proposal that the CC_r constitutes a major part of the neuroarchitecture of central systems. It will be recalled that centrencephalic theory generally views neocortical structures as playing a role in input processes and the CC_r as playing a role in central processes.

Visually Guided Responses and the CC_r

One of the more enduring and perplexing problems emerging from the neuropsychological analysis of visual discrimination learning in animals concerns the pathways beyond the primary cortical receiving areas that functionally relate the visual system with the motor system of the brain. The classical view that long transcortical pathways coursing between the occipital and the motor cortex mediate visually guided responding has been discredited to a large extent by neuroanatomical (Kuypers, Szwarcbart, Mishkin, & Rosvold, 1965; Myers, 1967), neurophysiological (Penfield, 1954a), and neuropsychological (Lashley, 1950; Myers, Sperry, & McCurdy, 1962; Passingham, 1987) data. Furthermore, it has long been known that neither the motor cortex (Lashley, 1950; Stepien, Stepien, & Konorski, 1961) nor the pyramidal tract (Lashley & Ball, 1929; Voneida, 1967) is essential for the maintenance of learned responses elicited (or guided) by optic impulses. Findings such as these led Lashley (1950) to conclude that "the conduction of impulses is from the retina to the lateral geniculate nuclei; thence to the striate areas, and from them down to some subcortical mechanism" (p. 467).

For a number of years, one of us (R.T.) attempted to identify this subcortical mechanism in the albino rat by determining which lesion placements abolished retention of a horizontal–vertical discrimination problem, while leaving intact retention of an equally difficult nonvisual inclined plane discrimination problem. This research effort yielded findings suggesting that there may be as many as three separate descending pathways from the occipital cortex which are significant in functionally linking the visual system with the motor system. These include an occipitostriatal (caudoputamenal) pathway (Thompson & Bachman, 1979a), an occipitoincertal pathway (Thompson & Bachman, 1979b), and an occipitoreticular pathway (Thompson & Peddy, 1979), the latter involving the region of the midbrain reticular formation (nucleus cuneiformis) immediately dorsolateral and caudal to the substantia nigra. What is remarkable about these findings is that the targets of these occipitofugal projection pathways lie in the vicinity of structures belonging to or anatomically associated with the CC_r (basal ganglia). In other words, the CC_r may constitute the subcortical mechanism to which Lashley referred in connection with the fate of impulses beyond the occipital cortex. These findings, however, are to be interpreted with caution since they have not been replicated in different laboratories nor do they appear to harmonize with those obtained by other workers set on tracking pathways from the visual cortex to the motor system. Mishkin and his associates, for example, have succeeded in identifying by the lesion method a circuit that presumably is implicated in visual recognition memory in the monkey (Mishkin, 1982; Mishkin & Appenzeller, 1987). This circuit begins at the primary visual (striate) cortex and extends through the peristriate areas to area TE of the temporal lobe. From there, the pathway continues to the amygdaloid complex and hippocampal formation and thence to the mediodorsal and anterior thalamic nuclei. The pathway beyond the mediodorsal and anterior thalamic regions has not been explicated, although Mishkin has suggested that the midline nuclei of the thalamus may also be involved.

The reason for this discrepancy between the rat and monkey findings over the circuits mediating visually guided behaviors is not entirely clear. Mishkin has offered the possibility that visual recognition memory is carried out by a different set of neural circuits from that involved in the memory of conventional two-choice visual discrimination habits (Mishkin, Malamut, & Bachevalier, 1984). This difference, however, may relate to the number of links in the chain between the occipital cortex and the motor system (CC_r) when it is considered that the intralaminar nuclei of the thalamus (to which the mediodorsal and anterior thalamic nuclei may be connected) project to the caudatoputamen (Heimer *et al.*, 1985). Consistent with this possibility are the results of an altogether different series of lesion studies on the monkey suggesting that the basal ganglia are pivotal in mediating visually guided behavior (Passingham, 1987).

Anticipatory Sets and the CC_r

In discussing the inadequacies of various theoretical attempts to explain the anatomical basis of learning in terms of new connections between sensory and motor centers, Sperry (1955) introduced a factor long recognized to play an essential role in the acquisition and expression of conditioned responses and problem-solving tasks. This factor is the "anticipatory set," "facilitory set," or "expectancy," which is often manifested in learning situations as a readiness to respond. For Sperry, this anticipatory set, which develops through training, exerts an organizing effect on the brain on both the sensory and motor sides that facilitates the release of the appropriate motor response by the relevant stimulus events. According to this view, training produces structural changes (engrams) from which an anticipatory set emerges (through higher-level cerebral activity such as insight) rather than structural changes from which a pathway between a stimulus and a response is reinforced. It is important to note that these central facilitory sets operate continually in most behavioral situations. As Sperry summarized, "It is by means of differential facilitory sets that the brain is able to function as many machines in one, setting and resetting itself dozens of times in the course of a day, now for one type of operation, now for another" and "a great deal of plasticity in vertebrate behavior, including that of conditioning, is made possible, not through structural remodeling of the fibre pathways, but through dynamic readjustments in the background of central facilitation" (Sperry, 1955, p. 43). Aside from mentioning the complexity and diffuseness of the neural circuits involved, Sperry offered no hints concerning the identification of the neuroanatomical mechanisms supporting anticipatory sets.

In our view, anticipatory sets can reasonably be linked to the activities of the CC_r for a number of reasons. First, the CC_r features a superabundance of anatomical connections throughout the neuraxis that could arouse particular patterns of central excitation and inhibition in the service of establishing a given behavioral state of readiness. Second, anticipatory sets are reflected electrophysiologically in the phenomenon of "assimilation of rhythms"—the appearance during the intertrial interval of brain activity normally evoked by the conditioned stimulus—and other "endogenous" processes (John, 1967, 1972). Of considerable interest is the fact that these assimilated rhythms seem to appear most reliably within the subcortical systems of the brain (see also Olds, Mink, & Best, 1969, and Ray, Mirsky, & Pragay, 1982). Similarly, the mapping of those units within the rat brain that exhibit "conditioned responses" (Disterhoft & Olds, 1972; Linesman & Olds, 1973; Olds *et al.*, 1972) has revealed that these learning units are found in several regions of the brain composing the CC_r, including the globus pallidus, ventral tegmental area, and pontine reticular formation. Third, research on regional changes in metabolism and blood flow in the human cerebral cortex has yielded data strongly indicating the existence of anticipatory sets—certain brain fields are tuned (activated) prior to the execution of a voluntary response and prior to the reception of sensory information (Roland, 1985). This differential tuning (and eventual recruitment) of cortical areas for the task at hand very likely involves the participation of the basal ganglia and thalamus (Mazziotta, Phelps, & Carson, 1984; Roland, Meyer, Shibasaki, *et al.*, 1982). Finally, since the cardinal feature of an anticipatory set is a predisposition to respond in a certain way to a given stimulus event, it would follow that such a predisposition would largely be achieved by brain structures, such as the basal ganglia, that serve an intermediary role in motor function (see Buchwald et al., 1975; Neafsey, Hull, & Buchwald, 1978). It should be recalled that the basal ganglia and anatomically associated nuclei largely compose the CC_r.

Engrams of Experience and the CC_r

Although memory traces have traditionally been viewed as resident within the cerebral cortex, the evidence is equivocal (Meyer, 1988; Thompson, 1983; R. F. Thompson, 1986). In fact, there is substantial documentation of subcortically induced memory traces through studies demonstrating learning in rats lacking virtually their entire cerebral cortex (Beattie, Gray, Rosenfield, *et al.*, 1978; Meyer, Yutzey, Dalby, & Meyer, 1968; Oakley, 1979, 1981, 1983a; Thompson, 1959). Whether or not all memory traces are located subcortically is debatable, but it is becoming increasingly clear from the clinical and experimental literature (Squire & Butters, 1984) that multiple engram mechanisms rather than a monolithic engram system inhabit the mammalian brain. Such a view is not incompatible with the notion expressed in Chapter 5 that problem-solving can be characterized in terms of the coactivation and sequencing of selective neural (cortical and subcortical) mechanisms by the CC_r. Accordingly, memory traces may be strategically located at a number of different cortical and subcortical sites. First, it may be assumed that the engram develops at those sites that connect the CC_r with the various specific neural mechanisms. As discussed earlier in connection with visually guided responding, numerous pathways unite the CC_r with the visual discrimination mechanism. In this case, the engram would be expected to be found at the site of these interconnecting pathways and would consequently take on a localized character even though it would not be confined to a single region. Alternatively, it may be assumed that the engram is formed in neural circuits residing within each neural mechanism as well as between neural mechanisms. If this is the case, then the engram would take on a diffuse character, particularly in relation to those habits supported by five or more mechanisms (maze and detour problems). Then there is the possibility that the engram is stored only within the subcortical components of those neural mechanisms supporting the habit.

It may not be necessary to subscribe fully to any one of these alternatives. As mentioned earlier, learning is not a unitary phenomenon, and the retention of most problem-solving tasks probably requires multiple sets of different engram systems. Richard Thompson and his associates (1984, for example, have presented compelling evidence for the distinctions among what they term "nonspecific trace systems" (related to learned motivation), "specific trace systems" (related to the acquisition of adaptive responses), and "cognitive trace systems" (related to procedural and declarative learning). Conceivably, each trace system may be stored in a different region of the brain. A somewhat different classification of engram systems has come from a lesion study showing that any given problem-solving activity can be fractionated into three separate habits

(Thompson, 1979). One (incentive habit) would be related to the initiation of the learned response leading to a goal object; the second (location habit) would be related to the location of the goal object or the correct route to the goal object; and the third (response habit) would be related to the general motor sequence necessary to gain access to the goal object. In terms of Penfield's theory, it is envisaged that the engrams underlying any given habit will be stored, at least in part, within the CC_r and probably within the specific neural mechanisms sustaining the habit, as well as at the site of the interconnecting pathways between these mechanisms. These and other possibilities concerning the locus of the engram have been discussed elsewhere (Thompson & Yu, 1987). In any case, the existence of certain engram systems within the interior parts of the brain seems assured. Establishing such a finding would provide remarkable support for centrencephalic theory and would correspondingly weaken corticocentric theories of brain function.

General Intelligence and the CC_r

The relationship between general intelligence (*g*) and the nonspecific mechanism (CC_r) has been discussed at length in Chapter 6 and will not be completely reviewed here. It is sufficient to say that general intelligence can be approached from a psychometric (statistical) viewpoint as well as from a biological (adaptive) viewpoint. While psychometric *g* appears to be the product of the sampling of a number of distinct abilities common to different cognitive tests (as theorized by Thomson, 1951, and Detterman, 1987), biological *g* is strictly a measure of the ability to solve a wide variety of cognitive tests [one possible index of Halstead's (1947) biological intelligence]. Since the former is a function of the complexity of the tests involved (e.g., detour and maze habits, which have the highest loadings on psychometric *g*, require the integrity of more specific neural mechanisms than discrimination habits, which have the lowest loadings on psychometric *g*), psychometric *g* can be likened to a measure of the depth (or complexity) of problem-solving ability. In contrast, biological *g* can be compared to a measure of the width (or pervasiveness) of problem-solving ability. Note that impaired psychometric *g* would impact mainly on the performance of complex problems (those involving numerous neural mechanisms), while impaired biological *g* would depress performance on simple (those involving few neural mechanisms) as well as complex problems. As shown in Chapter 6, the CC_r bears a stronger relationship to biological *g* than to psychometric *g*. Based on the factor loadings listed in Table 6.9, psychometric *g*, at least with respect to the rat, is largely associated with the neocortical (occipitotemporal and parietal) and limbic forebrain (posterior cingulate and posterolateral hypothalamic) regions of the brain. (It must be more than coincidence that these findings are congruent with current neuropsychological notions that view the cerebral cortex and limbic forebrain as the neural substrates of complex information processing.)

While centrencephalic theory can point to the CC_r as the neural substrate of biological *g*, can it logically point to the cerebral cortex and limbic forebrain as the neural underpinnings of psychometric *g*? We believe it can. Detterman (1987) has claimed that psychometric *g* constitutes "a finite set of independent abilities." If this is indeed the case, then these independent abilities can be compared to Fodor's input modules, which, according to centrencephalic theory, would be supported by cortical (and conceivably limbic forebrain) areas (see earlier discussion of "Modularity, Centrals Systems, and the CC_r"). Thus, what centrencephalic theory has contributed is a model in which the neural substrates of both biological *g* (which is essentially synonymous with "street-smartness") and psychometric *g* (comparable to academic intelligence) can be incorporated. This is especially significant when considering the fact that there are virtually no other neuroanatomical theories of general intelligence (MacPhail, 1982).

Subcortical Dementia and the CC_r

In recent years, there has been an explosion of clinical reports bearing on the concept of "subcortical dementia."

While some writers (Brown & Marsden, 1988; Mayeux, Stern, Rosen, & Benson, 1983; Whitehouse, 1986) have challenged the usefulness of the distinction between "cortical dementia" (e.g., Alzheimer's disease) and subcortical dementia (frequently linked to parkinsonism, Huntington's disease, and progressive supranuclear palsy), others agree that this dichotomy should be preserved (Albert, 1978; Benson, 1983; Brandt, Folstein, & Folstein, 1988; Cummings, 1986; Huber & Paulson, 1985). There is no need to dwell on the various controversial issues involved, except to mention that this dichotomy is of questionable value from a neuroanatomical point of view since cortical pathology usually coexists with subcortical pathology in both forms of dementia. On the other hand, there is some agreement that a clinical dichotomy should be maintained because the clinical manifestations of subcortical dementia are qualitatively different from those of cortical dementia (but see Brown & Marsden, 1988). Perhaps the strongest argument in favor of this dichotomy rests on the frequency with which individual cases present with "classical" clinical features of subcortical dementia together with a preponderance of subcortical pathology.

While the syndrome associated with cortical dementia involves a progressive deterioration of memory and intellectual functions to which is added aphasia, agnosia, acalculia, and apraxia, that associated with subcortical dementia may consist of "slowing of cognition, memory disturbances, difficulty with complex intellectual tasks such as strategy generation and problem-solving, visuospatial abnormalities, and disturbances of mood and affect" (Cummings, 1986, p. 682). Motor abnormalities, such as bradykinesia, dysarthria, ataxia, rigidity, and chorea, are usually prominent features of subcortical dementia since clinicopathological studies have frequently disclosed involvement of extrapyramidal motor structures. Furthermore, disturbances in motivation, attention, and arousal are not uncommon and have even been proposed to underlie many of the clinical manifestations of subcortical dementia (Cummings, 1986).

Subcortical dementia is readily explicable in terms of centrencephalic theory. In the first place, there appears to be a moderate degree of correspondence between some of the clinical features of subcortical dementia and the behavioral deficits predicted to follow partial lesions to the CC_r (see Table 7.2). In addition, there is a conspicuous overlap between those brainstem regions associated with subcortical dementia and those composing the CC_r. For example, the primary neuropathology in Huntington's disease includes the caudate nucleus and putamen (Bruyn, Bots, & Dom, 1979; Vonsattel, Myers, Stevens, *et al.*, 1985), that in Parkinson's disease includes the substantia nigra and ventral tegmental area (Gasper & Gray, 1984; Javoy-Agid & Agid, 1980; Jellinger, 1986), that in progressive supranuclear palsy includes the globus pallidus, superior colliculus, and brainstem reticular formation (Agid, Javoy-Agid, Ruberg, *et al.*, 1986; Steele, Richardson, & Olszewski, 1964), and that in Hallervorden–Spatz disease includes the globus pallidus and substantia nigra (Dooling, Schoene, & Richardson, 1974; Perry, Norman, Yong *et al.*, 1985). This parallelism is reinforced by reports that relatively focal lesions to components of the CC_r (as opposed to the more diffuse lesions associated with parkinsonism, Huntington's disease, and progressive supranuclear palsy) are sufficient to provoke moderate to severe impairments in cognitive functions in spite of the apparent absence of pathology to the cerebral cortex or hippocampus (see Table 7.2).

While this degree of concordance at the neuroanatomical level is unlikely to be due to chance, two caveats must be noted. Patients with subcortical dementia often suffer pathology to multiple brain sites (including the cerebral cortex), a condition that hampers efforts to correlate the severity of clinical defects to the degree of pathology to the interior parts of the brain. And not to be ignored are the isolated reports of cases that failed to manifest serious intellectual deficits despite having known lesions to such elements of the CC_r as the globus pallidus (Iizuka, Hirayama, & Maehara, 1984; Kessler, Schwechheimer, Reuther, & Born, 1984; Kimura, Hahn, & Barnett, 1987; LaPlane, Baulac, Widlocher, & Dubois, 1984), ventrolateral thalamus (Miyamoto, Bekku, Moriyama, & Tsuchida,

1985), and brainstem reticular formation (Iizuka *et al.*, 1984; Kessler *et al.*, 1984; Kimura *et al.*, 1987). These discrepancies, however, may not be as serious as they appear at first when contemplating the variability of clinical outcomes that must inevitably arise from individual differences not only in the locus, extent, momentum, and bilateral symmetry of lesions, but in the preinjury intelligence/education level of the patients as well (Grafman, Salazar, Weingartner, *et al.*, 1986).

Mental Retardation and the CC_r

Mental retardation is not a unitary condition, from either an etiological, neuropathological, or behavioral point of view. It may be due to genetic, infectious, or environmental factors, the brain abnormalities may be present at cortical or subcortical levels, and the degree of cognitive involvement may range from mild to profound. From such a perspective, it is obvious that a single animal model of mental retardation will not suffice in the overall search for the neurobiological bases of subnormal general intellectual functioning. While most animal models of mental retardation have customarily been patterned after certain etiological conditions presumed to be associated with mental retardation in humans (Meier, 1970), other approaches cannot be entirely excluded from consideration. It seems reasonable that any condition imposed on an animal during the developmental period that results in a generalized problem-solving impairment (the hallmark of mental retardation) has the potential to shed new light not only on the neurological, physiological, and biochemical correlates of intellectual disabilities, but on the nature of the cognitive deficits underlying these disabilities (see Archer, 1987; Thompson *et al.*, 1986).

Granting the foregoing considerations, our young rats with lesions to the CC_r appear to represent a brain-injured animal model of mental retardation. The validity of this brain-damaged animal model can be evaluated, at least in part, by the fulfillment of the following two criteria: (1) The problem-solving disorder in the animal model should be similar in some respects to that in the human condition, and (2) the brain lesions associated with the problem-solving disorder in the animal model should be similar in some respects to those associated with the problem-solving disorder in the human condition. In regard to our young rats with discrete lesions to the CC_r, the first criterion appears to be met with ease since one of the distinguishing features of mental retardation is a generalized problem-solving impairment (Denny, 1964; Miller, Hale, & Stevenson, 1968; Zeaman & House, 1967). Another correspondence at the behavioral level concerns the nature of the generalized problem-solving impairment. In recent years, research on the cognitive functioning of human retardates has shifted away from the study of defects in component processes, such as inhibition, recent memory, associative learning, and attention, and has focused instead on the study of defects in some superordinate ability, such as "executive" functioning that transcends the basic components of psychological processes (Borkowski, Peck, & Damberg, 1983; Brooks, Sperber, & McCauley, 1984). Campione and Brown (1978, 1984) have especially emphasized the distinction between knowledge (associative learning) and the use of knowledge (transfer of learning), claiming that the "hallmark of intelligence is the ability to generalize information from one situation to another, and that this ability in turn depends upon effective 'executive control' " (Campione & Brown, 1978, p. 279). Interestingly, we have likewise been unsuccessful in accounting for the generalized learning impairment exhibited by our young rats with CC_r lesions in terms of defects in component (inhibition, attention, and recent memory) processes (Thompson, Harmon, & Yu, 1985; Thompson *et al.*, 1986, 1989b). On the other hand, we have recently completed an experiment suggesting that these "retarded" brain-damaged rats are suffering from a defect in executive control (Thompson, Bjelajac, Fukui, *et al.*, 1989). Specifically, young rats with CC_r lesions, despite "knowing how" to dig, failed to use (transfer) this skill (an ability that depends on executive functioning) to solve a tunnel-digging detour problem that is readily solved by sham-operated control rats (see Fig. 8.1).

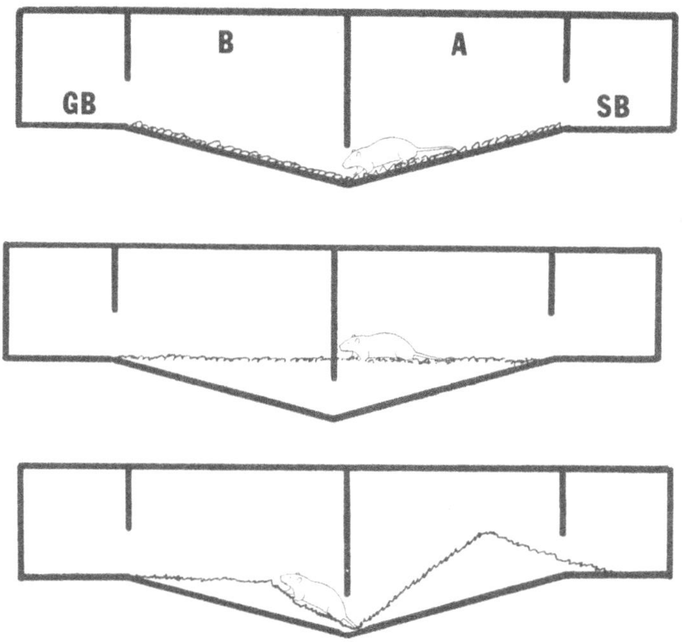

FIGURE 8.1. Schematic drawing of the tunnel-burrowing detour apparatus showing the start box (SB), two sections of the runway (A, B), and goal box (GB). The top panel shows a rat during a preliminary training trial, the middle panel shows a rat at the outset of the test trial, and the bottom panel shows a rat that has succeeded in burrowing through the sawdust to reach the goal box on the test trial.

Meeting the second criterion of an animal model of mental retardation is hindered by the fact that little is known about the parameters of the brain lesions that cause mental retardation in humans. However, in view of the diverse distribution of structural abnormalities reported by neuropathologists (Crome & Stern, 1972; Malamud, 1964; Shaw, 1987), it seems reasonable to assume that diffuse (and even possibly focal) lesions of considerable magnitude to either cortical or subcortical formations may underlie mental retardation. With respect to subcortical pathology, structural abnormalities somewhat restricted to the basal ganglia (Mito, Tanaka, Becker, *et al.*, 1986; Miyoshi, Matsuoka, & Mizushima, 1969; Towbin, 1969) and thalamus (Malamud, 1950) are not uncommon in the mentally retarded and severe intellectual impairments have been documented in adults with lesions involving the ventrolateral thalamus (Graff-Radford, Damasio, Yamada, *et al.*, 1985) and other brainstem regions at mesencephalic and pontine levels (Ueno & Takahata, 1978). And not to be ignored are those studies showing that there is a strong positive relationship between the severity of the intellectual deficit and the impairment in motor proficiency in the mentally retarded (Clausen, 1966; Distefano, Ellis, & Sloan, 1958; Heath, 1942; Hogg, 1982). This suggests that both the intellectual and motor disorders may result from the same pathological condition, which, in view of its pivotal role in motor function, may involve the basal ganglia. In any event, at this gross level of analysis, our rat data on the CC_r are compatible with the human data.

It should be noted that the foregoing treatment of our brainstem-injured animal model of mental retardation should not be interpreted as denying either the existence of a cortical basis of mental retardation or the usefulness of a cortical-injured animal model (see, for example, Archer, 1987). There now seems to be sufficient evidence that several pathobiological features of cortical neurons are associated with unclassified mental retardation (Purpura, 1979) and Down's syndrome (Scott, Becker, & Petit, 1983). Besides, centrencephalic theory would predict that diffuse pathology to the cerebral cortex alone could be a correlate of mental retardation since such a condition would compromise multiple input modules and lead to an impoverishment and derangement of information reaching the CC_r.

Thought Disorders and the CC_r

If the CC_r (which consists mainly of the basal ganglia and associated nuclei) is indeed concerned with higher mental functions, including thought processes, it might be expected that dysfunction of basal ganglia mechanisms would result not only in motor disorders, but in certain classes of neuropsychiatric disorders as well. According to Carlsson's (1988) recent speculations, this may in fact be the case. In Carlsson's view, psychosis may be due to overloading of the neocortex with sensory input and reticular activity as a consequence of increased tone of dopaminergic pathways. Under normal conditions, overloading of the neocortex is prevented by a corticostriatal–thalamocortical negative feedback loop; the more sensory input to the cortex, the more the striatal (caudate–putamen) complex is activated, which, in turn, activates the thalamic filter, resulting in reduced sensory input and reduced reticular activity to the cortex. This striatal–thalamic inhibition effect, however, can be compromised by increased activity of the nigrostriatal dopaminergic pathways which have their origin in the substantia nigra and adjacent ventral tegmental area.

Carlsson raises several interesting points in the development of the foregoing model which have a bearing on centrencephalic theory. First, he views the basal ganglia and related subcortical nuclei (particularly the caudate–putamen, globus pallidus, ventrolateral thalamus, substantia nigra, ventral tegmental area, and brainstem reticular formation) as controlling normal thought processes. These, of course, are the same structures we propose to make up the CC_r. Second, unlike other workers in the field who are inclined to believe that antipsychotic drugs act directly on the cerebral cortex, Carlsson reasons that the beneficial consequences of these drugs arise from the blocking of dopamine receptors in the striatal complex, which would have the effect of reducing overstimulation of the cortex through the increased filtering action of the thalamus. This notion, like centrencephalic theory as a whole, tends to deprive the neocortex of its supremacy in governing higher mental functions. And finally, Carlsson has argued from an evolutionary perspective that motor and cognitive skills had to have evolved in close conjunction and that this would also be true concerning their respective neuroanatomical substrates (the basal ganglia and cerebral cortex). Carlsson supports the close connection between motor and mental functions by referring to both animal studies on the disruption of conditioned responses by the administration of antidopaminergic drugs and clinical studies dealing with Huntington's disease and parkinsonism. Carlsson is not the only contemporary writer to stress the relationship between motor and mental functions. Hauert (1986), for example, presents evidence supporting the view that "motor function is a cognitive function." Ojemann (1982) has been more specific in stressing the interrelationships among those brain systems concerned with motor functions, on the one hand, and language, memory, and attention functions on the other. Similarly, in discussing the extrapyramidal system in general and the striatum in particular in their recent book on neuroanatomy, Nauta and Feirtag (1986) drew attention to the interrelationships among motor, ideational, and motivational factors and concluded with the following provocative question: "Is the striatum a place where brain mechanisms of movement, thought and motivation intersect?" (p. 312). These ideas concerning the relationship between cognition and motor function were, of course, proclaimed by Penfield three decades ago.

Some Complications

Penfield's centrencephalic theory is not without its flaws and shortcomings. Through the years, observations have been reported that are inimical to this theory. In addition, alternative accounts have been advanced of the very findings that Penfield viewed as compelling support for a centrencephalic-type theory. Besides, most neuropsychologists would deem it to be a step backward in neuroscience to refer to a functionally undifferentiated subcortical mechanism responsible for an array of behavioral and cognitive functions when considering that the alternative circumscribed modules of prefrontal lobes, posterior association cortex, limbic system, basal ganglia, brainstem reticular formation, and thalamocortical circuitry are available to explain the same phenomenon. The rest of this chapter is concerned with these issues.

Ascribing cognitive functions to a brainstem region based entirely on lesion data is a treacherous exercise to the extent that subcortical damage often causes dysfunctioning of brain tissue at cortical levels. In the case of subcortical dementia, which is usually associated with basal ganglia damage, several writers have suggested that the concomitant cognitive disturbances reflect dysfunctioning of the frontal cortex to which the basal ganglia are related (Freedman & Albert, 1985; Joynt & Shoulson, 1985; Pillon, Dubois, Lhermitte, & Agid, 1986; Taylor, Saint-Cyr, & Lang, 1986). The evidence for this suggestion rests not only on anatomofunctional relationships between the basal ganglia and frontal cortex (D'Antona, Baron, Samson, *et al.*, 1985; Leenders, Frackowiak, & Lees, 1988; Taylor *et al.*, 1986), but on similar syndromes arising from pathology to these two regions of the brain (Lichter, Corbett, Fitzgibbon, *et al.*, 1988; Pillon *et al.*, 1986; Taylor *et al.*, 1986). Thus, there exists a strong argument that lesions to the CC_r derange higher mental functions not so much as a result of injury to subcortical tissue, but because of the ensuing frontal lobe involvement.

This argument is hardly trifling and cannot readily be dismissed on clinical grounds. With respect to the rat data, however, this argument is discredited by the finding that while partial lesions to the basal ganglia (caudatoputamen, globus pallidus, or substantia nigra) produced a generalized learning impairment, extensive damage to certain sectors of the prefrontal cortex (more specifically, the frontocingulate and ventral frontal areas corresponding to the primate prefrontal cortex) did not (see Chapter 3). In view of the relatively poor development of the rodent's prefrontal cortex, these findings are admittedly equivocal. Nevertheless, these data deserve careful consideration in light of one of the conclusions drawn from a recent review of the comparative experimental literature on the frontal cortex—namely, "Thus, in spite of the tremendous differences in the relative volume of the frontal cortex of mammals, as well as the obvious diversity of behavioral repertories across mammalian phylogeny, there appears to be a remarkable unity in frontal cortex function across the class mammalia" (Kolb, 1984, p. 90).

Three additional points need to be mentioned in connection with the notion that cognitive deficits arising from CC_r lesions are accountable in terms of an attending dysfunction of the frontal cortex. First, there is evidence that a frontal lesion can impact on the activities of certain brainstem nuclei composing the CC_r. In one particular study involving measurements of regional differences in glucose metabolism in the monkey brain, the caudate, putamen, pallidum, ventrolateral thalamus, and superior colliculus manifested metabolic depression following unilateral lesions to the frontal association cortex (Deuel & Collins, 1983). Surprisingly, no changes were detected in those cortical areas known to have corticocortical connections with the lesion site. These data and others (Dauth *et al.*, 1985; Deuel & Collins, 1983; Hosokawa, Kato, Aiko, *et al.*, 1985) suggest that at least some of the behavioral disturbances arising from frontal lesions may be products of compromised basal ganglia function, due most likely to disruption of feedback loops between the frontal cortex and basal ganglia.

The second point concerns the prevailing view in contemporary brain research that cognitive functions are mediated almost exclusively by neocortical tissue, particularly the association areas. Clearly, this is the orientation that largely accounts for the tendency among writers not only to localize cognitive processes to the cerebral cortex, but to lose sight of the possible importance of brainstem mechanisms in higher intellectual activities. This view can be traced back to the classical theory of cortical function, which conceived of the sensory cortex as primarily playing a role in sensation and the association cortex as primarily playing a role in perception, including the formation of associations, the storage of memory traces, the implementation of problem-solving operations, and the translation of cognition into action (Diamond & Chow, 1962). When examined experimentally, however, the generality of this conceptualization is in doubt. For example, most of the so-called association cortex is neither multimodal nor "associational," and it now appears that the dominant function of the cerebral cortex is more related to sensory processing limited to single modalities (see Diamond, 1982; Kaas, 1987; Merzenich & Kaas, 1980). Furthermore, there seems to be little support for the idea that the association cortex (or neocortex as a whole) evolved to mediate learning or that it progressively appropriated this function from subcortical structures (MacPhail, 1982; Oakley, 1979, 1983a). Similarly, there is a conspicuous lack of definitive evidence that a specific cortical area or group of cortical areas serves as a storehouse of memory traces, despite the countless number of lesion studies undertaken to locate the engrams of experience (Lashley, 1950; Meyer & Meyer, 1982; Squire, 1987; Thompson, 1983; R. F. Thompson, 1986). This latter state of affairs particularly underscores the need to exercise caution in asserting that all cognitive processes are manifestations of neocortical activities. (See Meyer, 1988, for a particularly critical assessment of the claim that memory traces inhabit the cerebral cortex.)

Third, since the CC_r has a greater abundance of reciprocal connections with the frontal cortex than with any other cortical area (Heimer *et al.*, 1985), it is hardly surprising that selective lesions to these two brain regions would provoke similar patterns of behavioral deficits. In fact, it can be argued from a softened centrencephalic position that the anatomofunctional linkage between the CC_r and the frontal cortex represents evidence for a modest degree of corticalization of the CC_r. *In other words, the human equivalent of the* CC_r *may contain a neocortical component, namely, the prefrontal cortex.* This addendum to the CC_r in the case of the human brain accords with certain clinical findings that frontal cortical injuries tend to produce more dramatic deficits in problem-solving than more posteriorly localized cortical injuries (Goldstein & Levin, 1987; Shallice, 1982;

Vilkki, 1988). It is also compatible with Stuss and Benson's (1986) recent conceptualization of the prefrontal cortex: It "attends, integrates, formulates, executes, monitors, modifies, and judges all nervous system activities" (p. 248).

With respect to speech mechanisms, Penfield (1963) insisted that the centrencephalic system provided the network within which nonverbal concepts and word ideas were "hooked up." Accordingly, a degradation of language functions would be an anticipated consequence of a compromised CC_r. However, this prediction is clearly disconfirmed by the lack of language disturbances in cases of subcortical dementia (Cummings, 1986), although there are exceptions (Chui, Mortimer, Slager, & Webster, 1986; Matison, Mayeux, Rosen, & Fahn, 1982; Scott, Caird, & Williams, 1984). This discrepancy, while challenging to centrencephalic theory, may not be as devastating as it seems. In cases of subcortical dementia involving parkinsonism, Huntington's disease, progressive supranuclear palsy, and Hallervorden–Spatz syndrome, the underlying neuropathological changes tend to progress slowly. Conceivably, continual practice of language functions during this period of a slow-growing lesion could induce some kind of neural reorganization of speech mechanisms within the CC_r (Finger, 1978). This particular explanation is admittedly weak, but it fits well with the findings that sudden lesions to subcortical structures (such as the basal ganglia or thalamus) can precipitate language problems (Brunner, Kornhuber, Seemuller, *et al.*, 1982; Gorelick, Hier, Benevento, *et al.*, 1984; Graff-Radford *et al.*, 1985; Wallesch, Kornhuber, Brunner, *et al.*, 1983). Parenthetically, in line with the arguments of Fodor (1983), language may be more aptly conceptualized as an input module than as a central system, thus relieving the central systems (CC_r) of the burdens inherent in language comprehension and expression.

Shifting to another vein, Penfield's theory sustained a serious blow with the advent of the "split-brain" preparation. According to a strict centrencephalic position, which assumes that the activities of the cerebral hemispheres are coordinated and integrated through symmetrical connections of the brainstem, that these influences are exerted independently of the forebrain commissures, and that memory traces are located subcortically, it would be expected that animals subjected to sectioning of the forebrain commissures would be able to transfer complex learning from one hemisphere to the other and that high-level linguistic and visuospatial information would transfer from the left to the right hemisphere in patients with commissurotomy. In neither case, however, does this occur (Bogen, 1985; Gazzaniga & LeDoux, 1978; Myers, 1961; Sperry, 1961).

In response to these findings, centrencephalic theory will have to be revised accordingly to account for the fact that callosal fibers do function significantly under certain conditions to interrelate the activities of the cerebral hemispheres. But this fact alone does not invalidate the notion that the brainstem carries out additional coordinating and integrating functions. There are, for example, many instances of interhemispheric transfer of simple learned habits in split-brain animals and only a slight revision of centrencephalic theory is required to accommodate those findings pertaining to the absence of transfer of complex learned habits (Reader, 1974; Thompson, 1965). At the same time, it is becoming increasingly clear that there are many more demonstrations than previously supposed of interhemispheric exchange of information in complete commissurotomy patients (Cronin-Golomb, 1986; Sergent, 1987). And not to be ignored are the findings that split-brain humans behave as unified individuals (Cronin-Golomb, 1986; Sergent, 1987; Sperry, 1982). All of these examples suggest the existence of a subcortical bridge that functions in the cognitive domain. Parenthetically, the reports that patients with sectioned forebrain commissures may not evidence serious cognitive impairments (LeDoux, Risse, Springer, *et al.*, 1977; Zaidel, 1981) provide extraordinary support for Penfield's theory.

Some readers may recognize that most of the components of the CC_r strikingly coincide with those of the "R-complex" or reptilian forebrain, the most primitive of the three brains that MacLean (1975, 1980, 1985) theorizes that humans have inherited from their ancestors. (The remaining two include the paleomammalian brain, or limbic forebrain, and the neomammalian brain, or neocortex.) This R-complex is conceived to play a basic role in the automatic expression of genetically constituted, species-typical behaviors concerned with such activities as establishing a territory and social hierarchies, ritualistic displays involved in defense and courtship, foraging, hoarding, grooming, mating, migration, and the like. While this may indeed be the case (see Berntson & Micco, 1976), MacLean, unlike Penfield and ourselves, believes that the R-complex (or brainstem in general) is not endowed with the neural machinery necessary for learning. This notion is similar to the popular view that the basal ganglia are involved in the automatic execution of a sequence of motor programs, but are not involved in the actual learning or storage of these programs (Marsden, 1984). It also is in keeping with current neurophysiological views that only the limbic system and cerebral cortex are equipped with the necessary neurocircuitry to mediate learning and problem-solving. (It should be recalled, however, that the data included within Table 5.1 disclosed that such brainstem regions as the basal ganglia, limbic midbrain area, thalamus, and hypothalamus appear to be more concerned with overall problem-solving than the limbic forebrain and cerebral cortex.)

In reply to these notions, it must be pointed out that most learning that occurs in noncontrived natural settings is probably "automatic," requiring no special cognitive intervention such as rehearsal, choice of strategies, or reasoning. Also, learning is not an activity that we "learn"; rather we are programmed to learn just as we are programmed to indulge in certain nonlearned activities. In fact, there is evidence that animals are preprogrammed to learn particular things in particular ways (Gould & Marler, 1987). And last but not least, learning presupposes the prior existence of some species-typical behavior (e.g., an unconditioned response) and it can therefore be argued that learning is nothing more than the modification and elaboration of inherited response tendencies (Gallup, 1983). All of these considerations attest to the close relationship between genetically constituted behaviors and learned behaviors and consequently militate against simplistic attempts to segregate brain regions concerned with learning from those concerned with innately determined response tendencies. Besides, evidence abounds that learning can take place within a brain devoid of most of its cerebral cortex and limbic forebrain (Huston & Borbely, 1974; Huston, Joosten, & Tomaz, 1986; Huston, Tomaz, & Fix, 1985; Norman, Buchwald, & Villablanca, 1977) and there is even documentation for the occurrence of associative learning in spinal animals (Buerger & Fennessy, 1971; Chopin & Bennett, 1975; Sherman, Hoehler, & Buerger, 1982).

Finally, it is necessary to make a couple of comments on the critique of Penfield's theory by Walshe (1957). Many of the critical remarks made by Walshe were based (perhaps fairly) on the assumption that Penfield conceived of the centrencephalic integrating system as a "centrencephalon," a central locus acting independently of the surrounding brain mass (including the cortical gray mantle) and endowed with those higher-level functions related to consciousness, perception, motor planning, problem-solving, and memory. Undoubtedly, Penfield's emphasis in later writings (Penfield, 1958, 1966) that the centrencephalic system was a means of communication between upper and lower brain centers, that the multifarious roles played by this system emerged from functional interactions with other parts of the brain, and that the brainstem portion of this system was not to be likened to Descartes' pineal gland was a reply to Walshe's unrelenting commentary.

By defining Penfield's essential part of the centrencephalic system (CC_r) in terms of its participation in the acquisition of a broad spectrum of cognitively mediated tasks, it is possible to avoid a number of troublesome issues raised by Walshe. For example, the CC_r is no longer associated directly or indirectly with the seat of consciousness; it is now supported by observational data from both animals and humans, and its role in voluntary movement is enhanced by the finding that it is largely composed of basal ganglia structures.

9

Epilogue

In our view, the current experimental and clinical literature in neuropsychology and neurology is seriously losing sight of the importance of brainstem mechanisms in the mediation of higher mental functions. In an effort to offset this trend, an attempt has been made to restore some degree of luster and plausibility to Penfield's centrencephalic theory. While the fundamental impetus for this restoration grew out of newly acquired data on the nonspecific neural mechanism (formerly termed the "general learning system") in the laboratory rat, it was subsequently noted that recent clinical and pathological findings related especially to subcortical dementia converged on the same conclusion; namely, the basal ganglia and anatomically related nuclei constitute a brainstem mechanism that exercises a set of functions that seemingly intervenes in cognitively mediated behaviors. The precise set of functions that this brainstem ensemble is concerned with is, of course, problematic, but it could overlap with that attributed to the centrencephalic system by Penfield.

Probably most neuropsychologists do recognize the importance of brainstem mechanisms in higher mental functions, but in a way that does not infringe on or threaten the supremacy of the cerebral mantle in cognitive processes. Some may hold to the narrow view expressed by Luria (1970) that the main function of the brainstem in relation to cognition is to "energize" the neocortical areas. Others, however, may agree with the broader view articulated by Cummings (1986) that subcortical contributions involve "fundamental" functions, such as arousal, activation, attention, sequencing, motivation, and mood, as contrasted with the more cognitively loaded "instrumental" functions (language, praxis, perceptual recognition, memory, and calculation) subserved by neocortical regions. In any case, the homage paid to brainstem mechanisms as they bear on higher intellectual activities usually falls short of any suggestion that they actively participate in problem-solving and other cognitive processes. It is on this issue that Penfield's theory sharply diverges from other theories of brain function.

Testing Centrencephalic Theory

In some respects, centrencephalic-type theories yield no unique predictions beyond those derived from more conventional corticocentric-type theories of cerebral function. For example, both types of theories could readily explain the deleterious effects of focal or diffuse cortical injuries on learning, memory, and other intellectual skills (in the case of the former-type theories, input modules are disabled) and both would predict problem-solving impairments following brainstem injuries (in the case of the latter-type theories, fundamental functions are disabled). How, then, is it possible to falsify the particular centrencephalic theory adopted in this book? This question is by no means trivial when it is considered that scientists tend to gauge the usefulness of a theory in terms of the ease with which it can be tested.

The current theory literally stands or falls on the assumptions surrounding the functional significance ascribed to those structures composing the CC_r. By definition, bilateral lesions to the regions of the dorsal caudatoputamen, globus pallidus, ventrolateral thalamus, substantia nigra, ventral tegmental area, superior colliculus, median raphe, or pontine reticular formation would be expected to produce generalized problem-solving deficits in virtually all vertebrates that are members of

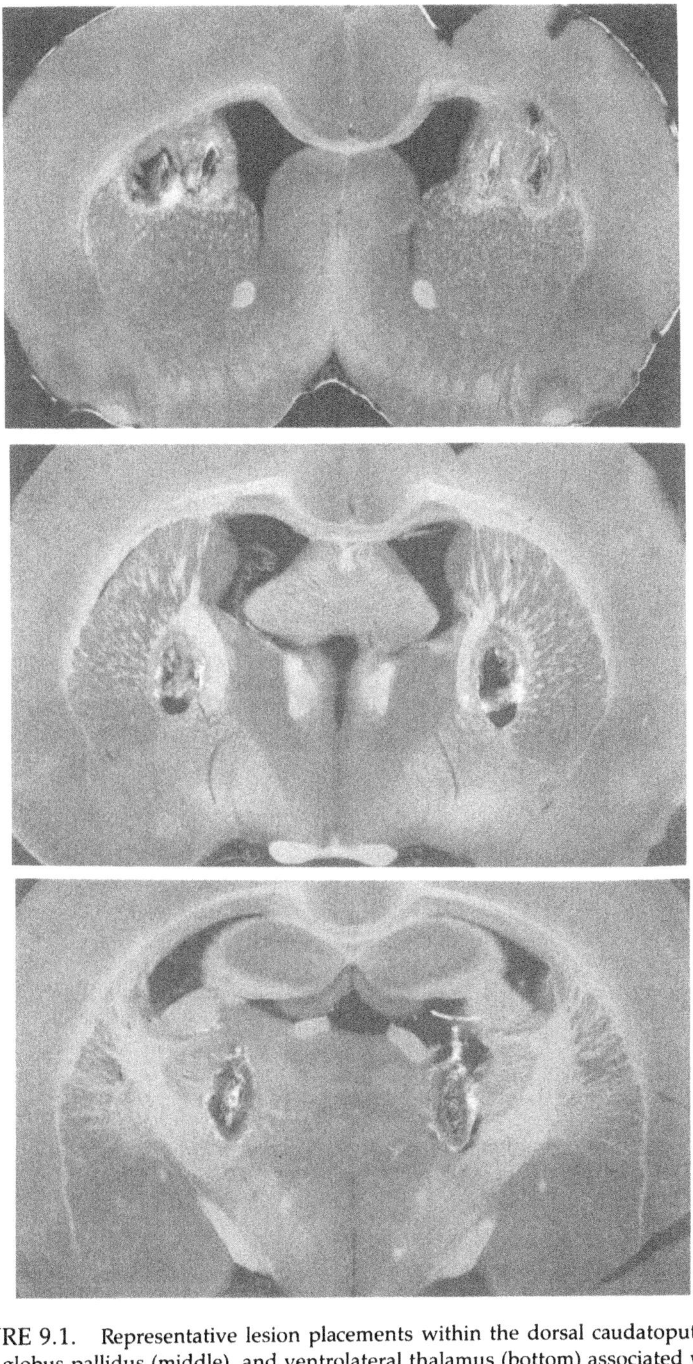

FIGURE 9.1. Representative lesion placements within the dorsal caudatoputamen (top), globus pallidus (middle), and ventrolateral thalamus (bottom) associated with a nonspecific problem-solving impairment in the rat.

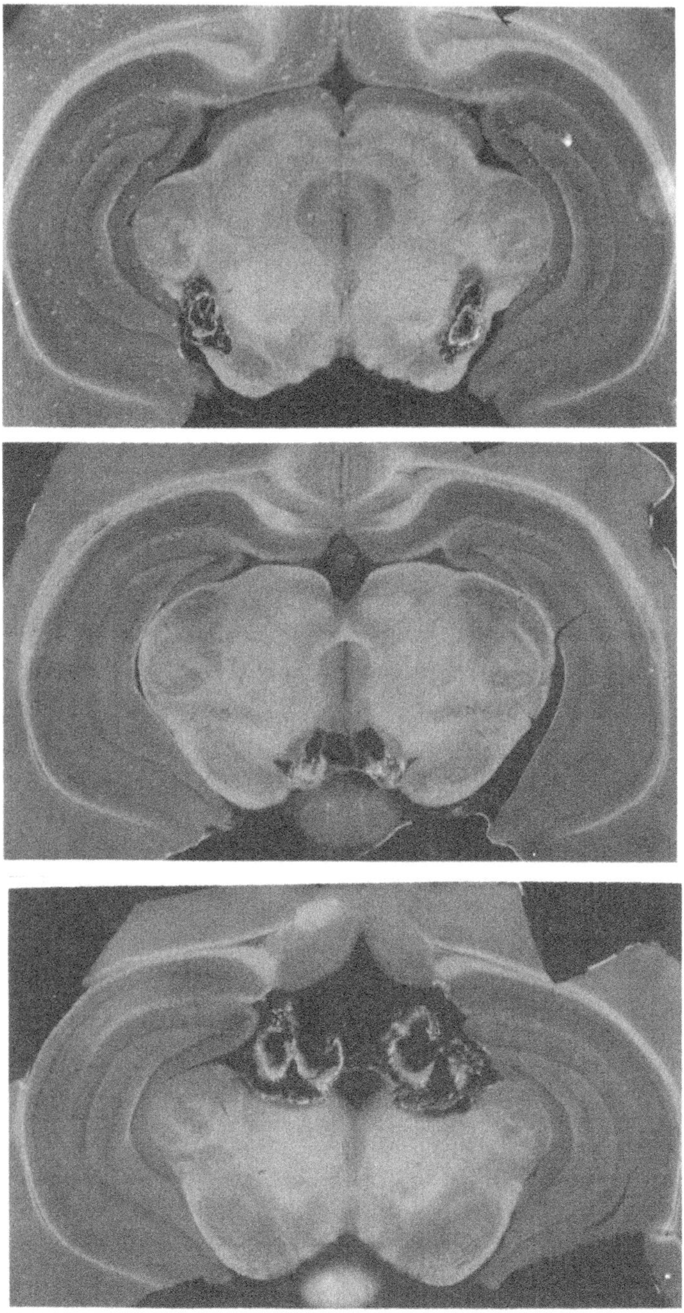

FIGURE 9.2. Representative lesion placements within the substantia nigra (top), ventral tegmental area (middle), and superior colliculus (bottom) associated with a nonspecific problem-solving impairment in the rat.

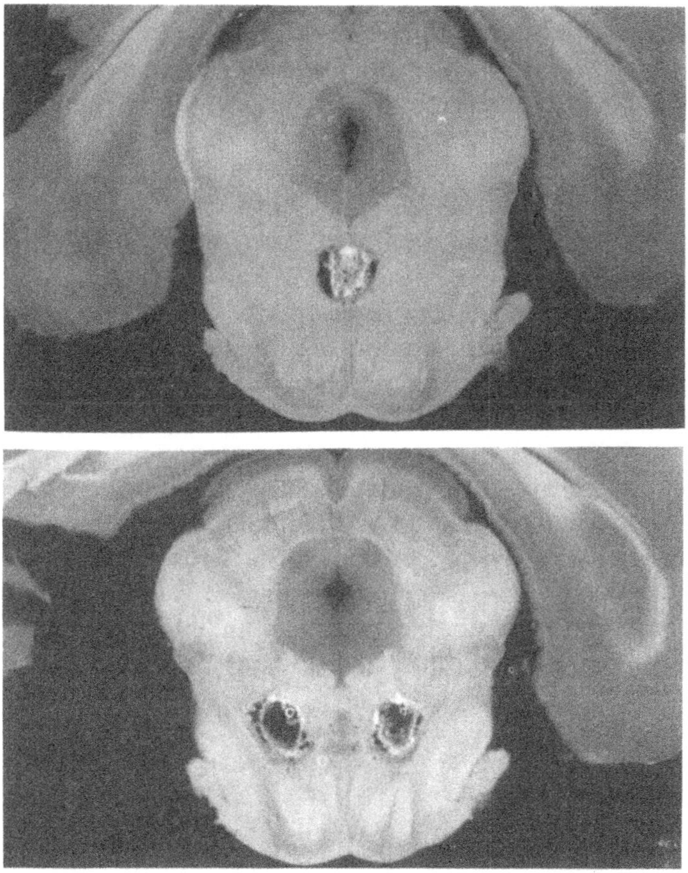

FIGURE 9.3. Representative lesion placements within the medial raphe (top) and pontine reticular formation (bottom) associated with a nonspecific problem-solving impairment in the rat.

the class Mammalia. (Because these brainstem structures are phylogenetically ancient relative to the neocortex, it would not be surprising if lesions to corresponding nuclei in fish, reptiles, and birds would likewise be associated with global problem-solving impairments.) Furthermore, these problem-solving deficits are not expected to be fully explicable in terms of motor, emotional, or motivational disorders nor are they expected to be satisfactorily reducible to a cortical involvement secondary to the brainstem lesion.

Obviously, a critical lesion experiment of the type mentioned above would require the production of moderate to extensive damage to any given component of the CC_r intended for investigation. For comparative purposes, samples of the individual electrolytic lesions to the CC_r that we investigated in the white rat are presented in Figures 9.1–9.3.

While other tests of the validity of centrencephalic theory are conceivable, they pale alongside the one concerned with focal lesions to the CC_r. If this theory is indeed correct, then only by replicating the generalized problem-solving deficit in different groups of laboratory animals will it be possible to moderate deeply rooted preconceptions about the sovereignty of the cerebral cortex in cognitive functions and to view future observations on brain–behavior relationships from a broader and less prejudiced conceptual orientation.

References

Agid, Y., Javoy-Agid, F., Ruberg, M., *et al.* (1986). Progressive supranuclear palsy: Anatomoclinical and biochemical considerations. *Advances in Neurology, 45,* 191–206.

Albert, M. L. (1978). Subcortical dementia. In R. Katzman, R. D. Terry, & K. L. Bick (Eds.), *Alzheimer's disease: Senile dementia and related disorders* (pp. 173–180). New York: Raven Press.

Ali-Cherif, A., Royere, M. L., Gosset, A., *et al.* (1984). Troubles du comportement et de l'activite mentale apres intoxication oxycarbonee. *Revue Neurologique, Paris, 140,* 401–405.

American Association on Mental Deficiency (1961). A manual on terminology and classification in mental retardation. *American Journal of Mental Deficiency,* pp. 1–15.

Andrezik, J. A., & Beitz, A. J. (1985). Reticular formation, central gray and related nuclei, In G. Paxinos (Ed.), *The rat nervous system. Vol. 2, Hindbrain and spinal cord* (pp. 1–28). New York: Academic Press.

Archer, T. (1987). Toward animal models of mental retardation. *Trends in Pharmacological Sciences, 8,* 165.

Barker, R. (1988). How does the brain control its own activity? A new function for the basal ganglia. *Journal of Theoretical Biology, 131,* 497–507.

Bayer, S. A. (1985). Hippocampal region. In G. Paxinos (Ed.), *The rat nervous system. Vol. 1. Forebrain and midbrain* (pp. 335–352). New York: Academic Press.

Beattie, M. S., Gray, T. S., Rosenfield, J. A., *et al.* (1978). Residual capacity for avoidance learning in decorticate rats: Enhancement of performance and demonstration of latent learning with *d*-amphetamine treatments. *Physiological Psychology, 6,* 279–287.

Beitz, A. J. (1987). The organization of afferent projections to the midbrain periaqueductal gray of the rat. *Neuroscience, 20,* 133–159.

Benson, D. F. (1983). Subcortical dementia: A clinical approach. In R. Mayeux & W. G. Rosen (Eds.), *The dementias* (pp. 185–193). New York: Raven Press.

Berlyne, D. E. (1970). Children's reasoning and thinking. In P. H. Mussen (Ed.), *Carmichael's handbook of developmental psychology, 3rd edition* (pp. 939–982). New York: Wiley.

Berntson, G. G., & Micco, D. J. (1976). Organization of brainstem behavioral systems. *Brain Research Bulletin, 1,* 471–483.

Bigl, V., Woolf, N. J., & Butcher, L. L. (1982). Cholinergic projections from the basal forebrain to frontal, parietal, temporal, occipital, and cingulate cortices: A combined fluorescent tracer and acetlycholinesterase analysis. *Brain Research Bulletin, 8,* 727–749.

Binet, A., & Simon, T. (1916). *The development of intelligence in children.* Baltimore: Williams & Wilkins.

Blakely, T. A., Crinella, F. M., & Fisher, T. A. (1985). Neuropsychological correlates of academic deficiency: Subtypes identified by object-cluster analysis. Paper presented at the annual meeting of the National Academy of Neuropsychologists, San Diego.

Bogen, J. E. (1985). Split-brain syndromes. In J. A. M. Frederiks, (Ed.), *Handbook of clinical neurology* (pp. 99–106). Amsterdam: Elsevier.

Borkowski, J. G., Peck, V. A., & Damberg, P. R. (1983). Attention, memory and cognition. In J. L. Matson & J. A. Mulick (Eds.), *Handbook of mental retardation* (pp. 479–498). New York: Pergamon Press.

Brandt, J., Folstein, S. E., & Folstein, M. F. (1988). Differential cognitive impairment in Alzheimer's disease and Huntington's disease. *Annals of Neurology, 23,* 555–561.

Brooks, P. H., Sperber, R., & McCauley, C. (1984). *Learning and cognition in the mentally retarded.* Hillsdale, NJ: Lawrence Erlbaum.

Brown, R. G., & Marsden, C. D. (1988). "Subcortical dementia": The neuropsychological evidence. *Neuroscience, 25,* 363–387.

Brunner, R. J., Kornhuber, H. H., Seemuller, E., *et al.* (1982). Basal ganglia participation in language pathology. *Brain and Language, 16,* 281–299.

Bruyn, G. W., Bots, G. T. A. M., & Dom, R. (1979). Huntington's chorea: Current neuropathological status. *Advances in Neurology, 23,* 83–93.

Buchwald, N. A., Hull, C. D., Levine, M. S., & Villablanca, J. (1975). The basal ganglia and the regulation of response and cognitive sets. In M. A. B. Brazier (Ed.), *Growth and development of the brain* (pp. 171–189). New York: Raven Press.

Buerger, A. A., & Fennessy, A. (1971). Long-term alteration of leg position due to shock avoidance by spinal rats. *Experimental Neurology, 30,* 195–211.

Campione, J. C., & Brown, A. L. (1978). Toward a theory of intelligence: Contributions from research with retarded children. *Intelligence, 2,* 279–304.

Campione, J. C., & Brown, A. L. (1984). Learning ability and transfer propensity as sources of individual differences in intelligence. In P. H. Brooks, R. Sperber, & C. McCauley (Eds.), *Learning and cognition in the mentally retarded* (pp. 265–294). Hillsdale, NJ: Lawrence Erlbaum.

Cardo, B. (1961). Rapports entre le niveau de vigilance et le conditionnement chez l'animal étude pharmacologique et neurologique. *Journal de Physiologie, Paris, 53,* 1–212.

Carlsson, A. (1988). Speculations on the control of mental and motor functions by dopamine-modulated cortico–striato–thalamo–cortical feedback loops. *The Mount Sinai Journal of Medicine, 55,* 6–10.

Carroll, J. B. (1988). Individual differences in cognitive functioning. In R. C. Atkinson, R. J. Herrnstein, G. Lindzey, & R. D. Luce (Eds.), *Steven's handbook of experimental psychology* (pp. 813–862). New York: Wiley Interscience.

Cattell, R. B. (1963). Theory of fluid and crystallized intelligence: A critical experiment. *Journal of Educational Psychology, 54,* 1–22.

Cattell, R. B. (1971). *Abilities: Their structure, growth, and action.* Boston: Houghton Mifflin.

Chopin, S. F., & Bennett, M. H. (1975). The effect of unavoidable shock on instrumental avoidance conditioning in spinal rats. *Physiology & Behavior, 14,* 399–401.

Chui, H. C., Mortimer, J. A., Slager, U., & Webster, D. D. (1986). Pathologic correlates of dementia in Parkinson's disease. *Archives of Neurology, 43,* 991–995.

Clausen, J. (1966). *Ability structure and subgroups in mental retardation.* London: Macmillan.

Commins, E. F., McNemar, Q., & Stone, C. P. (1932). Intercorrelations of measures of ability in the rat. *Journal of Comparative Psychology, 14,* 225–235.

Cools, A. R., Van Den Berken, J. H. L., Horstink, M. W. I., *et al.* (1984). Cognitive and motor shifting aptitude disorder in Parkinson's disease. *Journal of Neurology, Neurosurgery, & Psychiatry, 47,* 443–453.

Crinella, F. M. (1973). Identification of brain dysfunction syndromes in children through profile analysis: Patterns associated with so-called "minimal brain dysfunction." *Journal of Abnormal Psychology, 82,* 33–45.

Crinella, F. M., & Dreger, R. M. (1972). Tentative identification of brain dysfunction syndromes in children through profile analysis. *Journal of Consulting and Clinical Psychology, 38,* 251–260.

Crome, L., & Stern, J. (1972). *Pathology of mental retardation.* Baltimore: Williams & Wilkins.

Cronbach, L. J. (1957). Two disciplines of scientific psychology. *American Psychologist, 12,* 671–684.

Cronbach, L. J. (1975). Beyond the two disciplines of scientific psychology. *American Psychologist, 30,* 116–127.

Cronin-Golomb, A. (1986). Subcortical transfer of cognitive information in subjects with complete forebrain commissurotomy. *Cortex, 22,* 499–519.

Cummings, J. L. (1986). Subcortical dementia. *British Journal of Psychiatry, 149,* 682–697.

D'Antona, R., Baron, J. C., Samson, Y., *et al.* (1985). Subcortical dementia. Frontal cortex hypometabolism detected by positron tomography in patients with progressive supranuclear palsy. *Brain, 108,* 785–799.

Dauth, G. W., Gilman, S., Frey, K. A., *et al.* (1985). Basal ganglia glucose utilization after recent precentral ablation in the monkey. *Annals of Neurology, 17,* 431–438.

Davids, F. C., & Tolman, E. C. (1924). A note on the correlation between two mazes. *Journal of Comparative Psychology, 4,* 125–135.

Deiker, T., & Bruno, R. D. (1976). Sensory reinforcement of eyeblink rate in a decorticate human. *American Journal of Mental Deficiency, 80,* 665–667.

Delafresnaye, J. F. (1954). *Brain mechanisms and consciousness.* Springfield, IL: Charles C Thomas.

Denny, M. R. (1964). Research in learning and performance. In H. A. Stevens & R. Heber (Eds.), *Mental retardation* (pp. 100–142). Chicago: University of Chicago Press.

Denny-Brown, D. (1962). The midbrain and motor integration. *Proceedings of the Society for Biology and Medicine, 55,* 527–538.

Detterman, D. (1987). Theoretical notions of intelligence and mental retardation. *American Journal of Mental Deficiency, 92,* 2–11.

Deuel, R. K., & Collins, R. C. (1983). Recovery from unilateral neglect. *Experimental Neurology, 81,* 733–748.

Deuel, R. K., & Collins, R. C. (1984). The functional anatomy of frontal lobe neglect in the monkey: Behavioral and quantitative 2-deoxyglucose studies. *Annals of Neurology, 15,* 521–529.

Diamond, I. T. (1982). The functional significance of architectonic subdivisions of the cortex: Lashley's criticism of the traditional view. In J. Orbach (Ed.), *Neuropsychology after Lashley* (pp. 101–136). Hillsdale, NJ: Lawrence Erlbaum.

Diamond, I. T., & Chow, K. L. (1962). Biological psychology. In S. Koch (Ed.), *Psychology: A study of a science. Vol. 4* (pp. 158–241). New York: McGraw–Hill.

Diamond, S. J. (1980). *Neuropsychology.* Boston: Butterworths.

Distefano, M. K., Ellis, N. R., & Sloan, W. (1958). Motor proficiency in mental defectives. *Perceptual and Motor Skills, 8,* 231–234.

Disterhoft, J. F., & Olds, J. (1972). Differential development of conditioned unit changes in thalamus and cortex of rat. *Journal of Neurophysiology, 35,* 665–684.

Donoghue, J. P., Kerman, K. L., & Ebner, F. F. (1979). Evidence for two organizational plans within the somatic sensory–motor cortex of the rat. *Journal of Comparative Neurology, 183,* 647–664.

Dooling, E. C., Schoene, W. C., & Richardson, E. P. (1974). Hallervorden-Spatz syndrome. *Archives of Neurology, 30,* 70–83.

Douglas, R. J., Clark, G. M., Hubbard, D. G., & Wright, C. G. (1979). Effects of genetic vestibular defects on behavior related to spatial orientation and emotionality. *Journal of Comparative and Physiological Psychology, 93*, 467–480.

Eberhart, J. A., Morrell, J. I., Krieger, M. S., & Pfaff, D. W. (1985). An autoradiographic study of projections ascending from the midbrain central gray, and from the region lateral to it, in the rat. *Journal of Comparative Neurology, 241*, 285–310.

Evarts, E. V., & Thach, W. T. (1969). Motor mechanisms of the CNS: Cerebrocerebellar interrelations. *Annual Review of Physiology, 31*, 451–498.

Eysenck, H. J. (1961). Classification and the problem of diagnosis. In H. J. Eysenck (Ed.), *Handbook of abnormal psychology* (pp. 1–31). New York: Basic Books.

Eysenck, H. J. (1982a). The psychophysiology of intelligence. In C. D. Spielberger & J. N. Butcher (Eds.), *Advances in personality assessment* (pp. 1–33). Hillsdale, NJ: Lawrence Erlbaum.

Eysenck, H. J. (1982b). *A model for intelligence.* New York: Springer.

Eysenck, H. J., & Barrett, P. (1985). Psychophysiology and the measurement of intelligence. In C. R. Reynolds & V. L. Willson (Eds.), *Methodological and statistical advances in the study of individual differences* (pp. 1–49). New York: Plenum Press.

Fallon, J. H., & Laughlin, S. E. (1985). Substantia nigra. In: G. Paxinos (Ed.), *The rat nervous system. Vol. 1. Forebrain and midbrain* (pp. 353–374). New York: Academic Press.

Faull, R. L. M., & Carman, J. B. (1978). The cerebellofugal projections in the brachium conjunctivum of the rat. *Journal of Comparative Neurology, 178*, 495–518.

Faull, R. L. M., & Mehler, W. R. (1985). Thalamus. In G. Paxinos (Ed.), *The rat nervous system. Vol. 1. Forebrain and midbrain* (pp. 129–168). New York: Academic Press.

Faull, R. L. M., Nauta, W. J. H., & Domesick, V. B. (1986). The visual cortico–striato–nigral pathway in the rat. *Neuroscience, 19*, 1119–1132.

Finger, S. (1978). Lesion momentum and behavior. In S. Finger (Ed.), *Recovery from brain damage* (pp. 135–164). New York: Plenum Press.

Fodor, J. A. (1983). *The modularity of mind.* Cambridge, MA: MIT Press.

Fonberg, E. (1975). Improvement produced by lateral amygdala lesions on the instrumental alimentary performance impaired by dorsomedial amygdala lesions in dogs. *Physiology and Behavior, 14*, 711–717.

Frane, J. W., Jennrich, R. I., & Sampson, P. (1983). Factor analysis. In W. J. Dixon (Ed.), *BMDP statistical software* (pp. 480–508). Los Angeles: University of California Press.

Frank, J., & Levinson, H. N. (1973). Dysmetric dyslexia and dyspraxia: Hypothesis and study. *Journal of American Academy of Child Psychiatry, 12*, 690–701.

Freedman, M., & Albert, M. L. (1985). Subcortical dementia. In J. A. M.

Frederiks (Ed.), *Handbook of clinical neurology. Vol. 2* (pp. 311–316). Amsterdam: Elsevier.
Frommer, G. P. (1981). Tactile discrimination and somatosensory evoked responses after midbrain lesions in cats and rats. *Experimental Neurology, 73*, 755–800.
Gallup, G. G. (1983). Overcoming our resistance to animal research: Man in comparative perspective. In D. W. Rajecki (Ed.), *Comparing behavior: Studying man studying animals* (pp. 5–26). Hillsdale, NJ: Lawrence Erlbaum.
Gasper, P., & Gray, F. (1984). Dementia in idiopathic Parkinson's disease. *Acta Neuropathologica (Berlin), 64*, 43–52.
Gazzaniga, M. S., & LeDoux, J. E. (1978). *The integrated mind.* New York: Plenum Press.
Gerfen, C. R., Staines, W. A., Arbuthnott, G. W., & Fibiger, H. C. (1982). Cross connections of the substantia nigra in the rat. *Journal of Comparative Neurology, 207*, 283–303.
Giesler, G. J., Menetrey, D., & Basbaum, A. I. (1979). Differential origins of spinothalamic tract projections to medial and lateral thalamus in the rat. *Journal of Comparative Neurology, 184*, 107–126.
Goldberg, E., Antin, S. P., Bilder, R. M., *et al.* (1981). Possible role of mesencephalic reticular activation in long-term memory. *Science, 213*, 1392–1394.
Goldstein, F. C., & Levin, H. S. (1987). Disorders of reasoning and problem-solving ability. In M. J. Meier, A. L. Benton, & L. Diller (Eds.), *Neuropsychological rehabilitation* (pp. 327–354). New York: Guilford Press.
Goldstein, L. H., & Oakley, D. A. (1985). Expected and actual behavioral capacity after diffuse reduction in cerebral cortex: A review and suggestions for rehabilitative techniques with the mentally handicapped and head injured. *British Journal of Clinical Psychology, 24*, 13–24.
Gorelick, P. B., Hier, D. B., Benevento, L., *et al.* (1984). Aphasia after left thalamic infarction. *Archives of Neurology, 41*, 1296–1298.
Gorsuch, R. L. (1983). *Factor analysis, 2nd edition.* Hillsdale, NJ: Lawrence Erlbaum.
Gould, J. L., & Marler, P. (1987). Learning by instinct. *Scientific American, 256*, 74–85.
Graff-Radford, N. R., Damasio, H., Yamada, T., *et al.* (1985). Nonhaemorragic thalamic infarction. *Brain, 108*, 485–516.
Grafman, J., Salazar, A., Weingartner, H., *et al.* (1986). The relationship of brain-tissue loss volume and lesion location to cognitive deficit. *Journal of Neuroscience, 6*, 301–307.
Guilford, J. P. (1964). Zero correlations among tests of mental abilities. *Psychological Bulletin, 61*, 401–404.
Halstead, W. C. (1947). *Brain and intelligence.* Chicago: University of Chicago Press.

Halstead, W. C. (1951). Brain and intelligence. In L. A. Jeffries (Ed.), *Cerebral mechanisms in behavior* (pp. 244–272). New York: Wiley.
Harman, H. H. (1967). *Modern factor analysis, 2nd edition.* Chicago: University of Chicago Press.
Haroian, A. J., Massopust, L. C., & Young, P. A. (1981). Cerebello–thalamic projections in the rat: An autoradiographic and degeneration study. *Journal of Comparative Neurology, 197,* 217–236.
Hassler, R. (1977). Basal ganglia systems regulating mental activity. *International Journal of Neurology, 12,* 53–72.
Hassler, R. (1980). Brain mechanisms of intention and attention with introductory remarks on other volitional processes. In H. H. Kornhuber & L. Deecke (Eds.), *Motivation, motor and sensory processes of the brain* (pp. 585–614). Amsterdam: Elsevier.
Hauert, C. A. (1986). The relationship between motor function and cognition in the developmental perspective. *Italian Journal of Neurological Science, Supplement, 5,* 101–107.
Heath, S. R. (1942). Rail-walking performance as related to mental age and etiological type among the mentally retarded. *American Journal of Psychology, 55,* 240–247.
Hebb, D. O. (1945). Man's frontal lobe. *Archives of Neurology and Psychiatry, 54,* 10–24.
Heber, R. (1962). Mental retardation: Concept and classification. In P. E. Trapp & P. Himelstein (Eds.), *Readings on the exceptional child* (pp. 28–45). New York: Appleton–Century–Crofts.
Heilman, K. M., Watson, R. T., Valenstein, E., & Goldberg, M. E. (1987). Attention: Behavior and neural mechanisms. In F. Plum (Ed.), *Handbook of physiology. Section I. The nervous system. Vol. V. Higher functions of the brain* (pp. 461–481). Bethesda, MD: American Physiological Society.
Heimer, L., Alheid, G. F., & Zaborszky, L. (1985). Basal ganglia. In G. Paxinos (Ed.), *The rat nervous system. Vol. 1. Forebrain and midbrain* (pp. 37–86). New York: Academic Press.
Herkenham, M. (1986). New perspectives on the organization and evolution of nonspecific thalamocortical projections. In E. G. Jones (Ed.), *Cerebral cortex. Vol. 5* (pp. 403–445). New York: Plenum Press.
Herkenham, M., & Nauta, W. J. H. (1979). Efferent connections of the habenular nuclei in the rat. *Journal of Comparative Neurology, 187,* 19–48.
Hocherman, S., Aharonson, D., Medalion, & Hocherman, I. (1988). Perception of the immediate extrapersonal space through proprioceptive inputs. *Experimental Brain Research, 73,* 256–262.
Hogg, J. (1982). Motor development and performance of severely mentally handicapped people. *Developmental Medicine and Child Neurology, 24,* 188–193.
Hosokawa, S., Kato, M., Aiko, Y., *et al.* (1985). Altered local cerebral glucose

utilization by unilateral frontal cortical ablation in rats. *Brain Research, 343*, 8–15.

Huber, S. J., & Paulson, G. W. (1985). The concept of subcortical dementia. *American Journal of Psychiatry, 141*, 1312–1317.

Huston, J. P., & Borbely, A. A. (1974). The thalamic rat: General behavior, operant learning with rewarding hypothalamic stimulation, and effects of amphetamine. *Physiology & Behavior, 12*, 433–448.

Huston, J. P., Joosten, M., & Tomaz, C. (1986). Reversal learning of an avoidance response in detelencephalated rats. *Experimental Neurology, 91*, 147–153.

Huston, J. P., Tomaz, C., & Fix, I. (1985). Avoidance learning in rats devoid of the telencephalon plus thalamus. *Behavioral Brain Research, 17*, 87–95.

Iizuka, R., Hirayama, K., & Maehara, K. (1984). Dentato–rubro–pallido–luysian atrophy: A clinico–pathological study. *Journal of Neurology, Neurosurgery and Psychiatry, 47*, 1288–1298.

Irle, E. (1985). Combined lesions of septum, amygdala, hippocampus, anterior thalamus, mammillary bodies and cingulate and subicular cortex fail to impair the acquisition of complex learning tasks. *Experimental Brain Research, 58*, 346–361.

Irle, E., & Markowitsch, H. J. (1983). Differential effects of double and triple lesions of the cat's limbic system on subsequent learning behavior. *Behavioral Neuroscience, 97*, 908–920.

Irle, E., & Markowitsch, H. J. (1984). Differential effects of prefrontal lesions and combined prefrontal and limbic lesions on subsequent learning performance in the cat. *Behavioral Neuroscience, 98*, 884–897.

Isaacson, R. L. (1976). Experimental brain lesions and memory. In M. R. Rosenzweig & E. L. Bennett (Eds.). *Neural mechanisms of learning and memory* (pp. 521–543). Cambridge, MA: MIT Press.

Iversen, S. D. (1973). Brain lesions and memory in animals. In J. A. Deutsch (Ed.). *The physiological basis of memory* (pp. 305–364). New York: Academic Press.

Izard, C. E. (1971). *The face of emotion*. New York: Appleton–Century–Crofts.

Jasper, H. H. (1977). The centrencephalic system. *Canadian Medical Association Journal, 116*, 1371–1372.

Javoy-Agid, F., & Agid, Y. (1980). Is the mesocortical dopaminergic system involved in Parkinson disease? *Neurology, 30*, 1326–1330.

Jellinger, K. (1986). Overview of morphological changes in Parkinson's disease. *Advances in Neurology, 45*, 1–18.

Jensen, A. R. (1980). *Bias in mental testing*. New York: Free Press.

Jensen, A. R. (1985). The nature of the black–white difference on various psychometric tests: Spearman's hypothesis. *Behavioral and Brain Sciences, 8*, 193–263.

John, E. R. (1967). *Mechanisms of memory*. New York: Academic Press.

John, E. R. (1972). Switchboard versus statistical theories of learning and memory. *Science, 177*, 850–864.

Johnston, M. V., McKinney, M., & Coyle, J. T. (1981). Neocortical cholinergic innervation: A description of extrinsic and intrinsic components in the rat. *Experimental Brain Research, 43,* 159–172.

Jones, B. E., & Yang, T. Z. (1985). The efferent projections from the reticular formation and the locus coeruleus studied by anterograde and retrograde axonal transport in the rat. *Journal of Comparative Neurology, 242,* 56–92.

Jones, E. G. (1985). *The thalamus.* New York: Plenum Press.

Joynt, R. J., & Shoulson, I. (1985). Dementia. In K. M. Heilman & E. Valenstein (Eds.), *Clinical neuropsychology* (pp. 453–479). New York: Oxford University Press.

Jung, R., & Hassler, R. (1960). The extrapyramidal motor system. In J. Field (Ed.), *Handbook of physiology. Section I, Vol. II* (pp. 863–927). Washington, DC: American Physiological Society.

Kaas, J. H. (1987). The organization of neocortex in mammals: Implications for theories of brain function. *Annual Review of Psychology, 38,* 129–151.

Keating, D. P. (1984). The emperor's new clothes: The "new look" in intelligence research. In R. J. Sternberg (Ed.), *Advances in the psychology of human intelligence, Vol. II* (pp. 1–45). Hillsdale, NJ: Lawrence Erlbaum.

Kelly, D. (1985). Central representations of pain and analgesia. In E. R. Kandel & J. H. Schwartz (Eds.), *Principles of neural science* (pp. 331–343). New York: Elsevier.

Kessler, C., Schwechheimer, K., Reuther, R., & Born, J. A. (1984). Hallervorden–Spatz syndrome restricted to the pallidal nuclei. *Journal of Neurology, 231,* 112–116.

Kilmer, W. L., McCulloch, W. S., & Blum, J. (1968). An embodiment of some vertebrate command and control principles. *Currents in Modern Biology, 2,* 81–97.

Kimura, D., Hahn, A., & Barnett, H. J. M. (1987). Attentional and perseverative impairment in two cases of familial fatal parkinsonism with cortical sparing. *Canadian Journal of Neurological Science, 14,* 597–599.

Kita, H., & Omura, Y. (1982). An anterograde HRP study of retinal projections to the hypothalamus in the rat. *Brain Research Bulletin, 8,* 249–253.

Kitai, S. T. (1981). Electrophysiology of the corpus striatum and brain stem integrating systems. In J. M. Brookhard & V. Mountcastle (Eds.), *Handbook of physiology. Section 1. Vol. II* (pp. 997–1015). Bethesda, MD: American Physiological Society.

Klopf, A. H. (1982). *The hedonistic neuron.* New York: Hemisphere.

Knoll, J. (1969). *The theory of active reflexes.* New York: Hafner.

Knook, H. L. (1965). *The fibre-connections of the forebrain.* Philadelphia: F. A. Davis.

Kohen-Raz, R. (1986). *Learning disabilities and postural control.* London: Freund.

Kohler, W. (1925). *The mentality of apes.* New York: Harcourt Brace.

Kolb, B. (1984). Functions of the frontal cortex of the rat: A comparative review. *Brain Research Review, 8,* 65–98.

Kolb, B., & Whishaw, I. Q. (1988). Mass action and equipotentiality reconsidered. In S. Finger, T. E. LeVere, C. R. Almli, & D. G. Stein (Eds.), *Brain injury and recovery* (pp. 103–116). New York: Plenum Press.

Konorski, J. (1967). *Integrative activity of the brain.* Chicago: University of Chicago Press.

Kornblith, C., & Olds, J. (1973). Unit activity in brain stem reticular formation of the rat during learning. *Journal of Neurophysiology, 36,* 489–501.

Kuypers, H. G. J. M., Szwarcbart, M. K., Mishkin, M., & Rosvold, H. E. (1965). Occipito-temporal corticocortical connections in the rhesus monkey. *Experimental Neurology, 11,* 245–262.

LaPlane, D., Baulac, M., Widlocher, D., & Dubois, B. (1984). Pure psychic akinesia with bilateral lesions of basal ganglia. *Journal of Neurology, Neurosurgery and Psychiatry, 47,* 377–385.

Lashley, K. S. (1929). *Brain mechanisms and intelligence.* Chicago: University of Chicago Press.

Lashley, K. S. (1943). Studies of cerebral function in learning. XII. Loss of the maze habit after occipital lesions in blind rats. *Journal of Comparative Neurology, 79,* 431–462.

Lashley, K. S. (1950). In search of the engram. *Symposium of the Society for Experimental Biology, 4,* 454–482.

Lashley, K. S., & Ball, J. (1929). Spinal conduction and kinesthetic sensitivity in the maze habit. *Journal of Comparative Psychology, 9,* 71–105.

LeDoux, J. E., Risse, G. L., Springer, S. P., *et al.* (1977). Cognition and commissurotomy. *Brain, 100,* 87–104.

Leenders, K. L., Frackowiak, R. S. J., & Lees, A. J. (1988). Steele–Richardson–Olszewski syndrome. *Brain, 111,* 615–630.

Legg, C. R. (1979). An autoradiographic study of the efferent projections of the ventral lateral geniculate nucleus of the hooded rat. *Brain Research, 170,* 349–352.

Leland, H., Shellhaas, M., Nihara, K., & Foster, R. (1967). Adaptive behavior: A new dimension in the classification of the mentally retarded. *Mental Retardation Abstracts, 4,* 359–387.

LeVere, T. E. (1975). Neural stability, sparing, and behavioral recovery following brain damage. *Psychological Review, 82,* 344–358.

Levinson, H. N. (1986). *Phobia free.* New York: M. Evans.

Levinson. H. N. (1988). The cerebellar–vestibular basis of learning disabilities in children, adolescents and adults: Hypothesis and study. *Perceptual and Motor Skills, 67,* 983–1006.

Lhermitte, F., & Signoret, J. L. (1976). The amnesic syndromes and the hippocampal–mamillary system. In M. R. Rosenzweig & E. L. Bennett (Eds.), *Neural mechanisms of learning and memory* (pp. 49–56). Cambridge, MA: MIT Press.

Lichter, D. G., Corbett, A. J., Fitzgibbon, G. M., *et al.* (1988). Cognitive and motor dysfunction in Parkinson's disease. *Archives of Neurology, 45,* 854–860.

Lidsky, T. I., Manetto, C., & Schneider, J. S. (1985). A consideration of sensory factors involved in motor functions of the basal ganglia. *Brain Research Review, 9,* 133–146.

Linesman, M. A., & Olds, J. (1973). Activity changes in rat hypothalamus, preoptic area, and striatum associated with Pavlovian conditioning. *Journal of Neurophysiology, 36,* 1038–1050.

Livesey, P. J. (1970). Consideration of the neural basis of intelligent behavior: Comparative studies. *Behavioral Science, 15,* 164–170.

Loughlin, S. E., & Fallon, J. H. (1985). Locus coeruleus. In G. Paxinos (Ed.), *The rat nervous system. Vol. 2, Hindbrain and spinal cord* (pp. 79–93). New York: Academic Press.

Lubar, J. F., Shostal, C. J., & Perachio, A. A. (1967). Nonvisual functions of visual cortex in the cat. *Physiology & Behavior, 2,* 179–184.

Luria, A. R. (1966). *Higher cortical functions in man.* New York: Basic Books.

Luria, A. R. (1970). The functional organization of the brain. *Scientific American, 222,* 66–78.

Lynch, G. (1976). Some difficulties associated with the use of lesion techniques in the study of memory. In M. R. Rosenzweig & E. L. Bennett (Eds.), *Neural mechanisms of learning and memory* (pp. 544–548). Cambridge, MA: MIT Press.

MacKay, D. M. (1966). Conscious control of action. In J. C. Eccles (Ed.), *Brain and conscious experience* (pp. 422–440). New York: Springer-Verlag.

Mackel, R., & Noda, T. (1989). The pretectum as a site for relaying dorsal column input to thalamic VL neurons. *Brain Research, 476,* 135–139.

MacLean, P. D. (1975). The evolution of three mentalities. *Man–Environment Systems, 5,* 213–224.

MacLean, P. D. (1980). On the origin and progressive evolution of the triune brain. In E. Armstrong & D. Falk (Eds.), *Primate brain evolution* (pp. 291–316). New York: Plenum Press.

MacLean, P. D. (1985). Brain evolution relating to family, play, and the separation call. *Archives of General Psychiatry, 42,* 405–417.

MacPhail, E. M. (1982). *Brain and intelligence in vertebrates.* Oxford: Clarendon Press.

Maier, N. R. F. (1936). Reasoning in children. *Journal of Comparative Psychology, 21,* 357–366.

Malamud, N. (1950). Status marmoratus: A form of cerebral palsy following either birth injury or inflammation of the central nervous system. *Journal of Pediatrics, 37,* 610–619.

Malamud, N. (1964). Neuropathology. In H. A. Stevens & R. Heber (Eds.), *Mental retardation* (pp. 429–452). Chicago: University of Chicago Press.

Marsden, C. D. (1984). Function of the basal ganglia as revealed by cognitive and motor disorders in Parkinson's disease. *Canadian Journal of Neurological Sciences, 11,* 129–135.

Marshall, J. C. (1984). Multiple perspectives on modularity. *Cognition, 17,* 209–242.

Massopust, L. C. (1961). Stereotaxic atlases: A. Diecephalon of the rat. In D. E. Sheer (Ed.), *Electrical stimulation of the brain* (pp. 182–202). Austin: University of Texas Press.

Matison, R., Mayeux,R., Rosen, J., & Fahn, S. (1982). "Tip-of-the-tongue" phenomenon in Parkinson disease. *Neurology, 32,* 567–570.

Matute, C., & Streit, P. (1985). Selective retrograde labeling with D-[^{3}H]-aspartate in afferents to the mammalian superior colliculus. *Journal of Comparative Neurology, 241,* 34–49.

Mayeux, R., Stern, Y., Rosen, J., & Benson, D. F. (1983). Is "subcortical dementia" a recognizable clinical entity? *Annals of Neurology, 14,* 278–283.

Mazziotta, J. C., Phelps, M. E., & Carson, R. E. (1984). Tomographic mapping of human cerebral metabolism: Subcortical responses to auditory and visual stimulation. *Neurology 34,* 825–828.

McCullock, T. C. (1935). A study of cognitive abilities of the white rat with special reference to Spearman's theory of two factors. *Contributions to Psychological Theory, 1,* 66.

McGeer, P. L., McGeer, E. G., Itagaki, S., & Mizukawa, K. (1987). Anatomy and pathology of the basal ganglia. *Canadian Journal of Neurological Sciences, 14,* 363–372.

McNew, J. J., & Thompson, R. (1966). Role of the limbic system in active and passive avoidance conditioning in the rat. *Journal of Comparative and Physiological Psychology, 61,* 173–180.

Mehler, W. R., & Rubertone, J. A. (1985). Anatomy of the vestibular nucleus complex. In G. Paxinos (Ed.), *The rat nervous system. Vol. 2, Hindbrain and spinal cord* (pp. 185–219). New York: Academic Press.

Meier, G. W. (1970). Mental retardation in animals. *International Review of Research in Mental Retardation, 4,* 263–310.

Meller, S. T., & Dennis, B. J. (1986). Afferent projections to the periaqueductal gray in the rabbit. *Neuroscience, 19,* 917–964.

Merzenich, M. M., & Kaas, J. H. (1980). Principles of organization of sensory-perceptual systems in mammals. *Progress in Psychobiology and Physiological Psychology, 9,* 1–42.

Meyer, D. R. (1988). Bases of inductions of recoveries and protections from amnesia. In S. Finger, T. E. LeVere, C. R. Almi, & D. G. Stein (Eds.), *Brain injury and recovery* (pp. 29–44). New York: Plenum Press.

Meyer, P. M., Johnson, D. A., & Vaughn, D. W. (1970). The consequences of septal and neocortical ablations upon learning a two-way conditioned avoidance response. *Brain Research, 22,* 113–120.

Meyer, P. M., & Meyer, D. R. (1982). Memory, remembering, and amnesia. In R. L. Isaacson & N. E. Spear (Eds.), *The expression of knowledge* (pp. 179–212). New York: Plenum Press.

Meyer, P. M., Yutzey, D. A., Dalby, D. A., & Meyer, D. R. (1968). Effects of simultaneous septal–visual, septal–anterior, and anterior–posterior lesions upon relearning a black–white discrimination. *Brain Research, 8,* 281–290.

Miller, L. K., Hale, G. A., & Stevenson, H. W. (1968). Learning and problem solving by retarded and normal Ss. *American Journal of Mental Deficiency, 72,* 681–690.

Mishkin, M. (1982). A memory system in the monkey. *Philosophical Transactions of the Royal Society, London, 298,* 85–95.

Mishkin, M., & Appenzeller, T. (1987). The anatomy of memory. *Scientific American, 256,* 80–89.

Mishkin, M., Malamut, B., & Bachevalier, J. (1984). Memories and habits: Two neural systems. In G. Lynch, J. L. McGaugh, & N. M. Weinberger (Eds.), *Neurobiology of learning and memory* (pp. 65–77). New York: Guilford Press.

Mito, T., Tanaka, T., Becker, L. E., *et al.* (1986). Infantile bilateral striatal necrosis. *Archives of Neurology, 43,* 677–680.

Miyamoto, T., Bekku, H., Moriyama, E., & Tsuchida, T. (1985). Present role of stereotactic thalamotomy for parkinsonism. *Applied Neurophysiology, 48,* 294–304.

Miyoshi, K., Matsuoka, T., & Mizushima, S. (1969). Familial holotopistic striatal necrosis. *Acta Neuropathologica, Berlin, 13,* 240–249.

Molinari, M., Minciacchi, D., & Bentivoglio, M. (1985). Efferent fibers from the motor cortex terminate bilaterally in the thalamus of rats and cats. *Experimental Brain Research, 57,* 305–312.

Mowrer, O. H. (1960). *Learning theory and behavior.* New York: Wiley.

Myers, R. E. (1961). Corpus callosum and visual gnosis. In J. F. Delafresnaye (Ed.), *Brain mechanisms and learning* (pp. 481–503). Oxford: Blackwell.

Myers, R. E. (1967). Cerebral connectionism and brain function. In C. H. Millikan & F. L. Darley (Eds.), *Brain mechanisms underlying speech and language* (pp. 61–72). New York: Grune & Stratton.

Myers, R. E., Sperry, R. W., & McCurdy, N. M. (1962). Neural mechanisms of visual guidance of limb movement. *Archives of Neurology 7,* 195–202.

Nagata, S. (1986). The vestibulothalamic connections in the rat: A morphological analysis using wheat germ agglutinin–horseradish peroxidase. *Brain Research, 376,* 57–70.

Nauta, W. J. H., & Domesick, V. B. (1981). Ramifications of the limbic system. In S. Matthysse (Ed.), *Psychiatry and the biology of the human brain* (pp. 165–188). New York: Elsevier/North-Holland.

Nauta, W. J. H., & Feirtag, M. (1986). *Fundamental neuroanatomy.* New York: W. H. Freeman.

Neafsey, E. J., Hull, C. D., & Buchwald, N. A. (1978). Preparation for movement in the cat. II. Unit activity in the basal ganglia and thalamus. *Electroencephalography and Clinical Neurophysiology 44,* 714–723.

Norman, R. J., Buchwald, J. S., & Villablanca, J. R. (1977). Classical conditioning with auditory discrimination of the eye blink in decerebrate cats. *Science 196,* 551–553.

Nothias, F., Peschanski, & Besson, J-M. (1988). Somatotopic reciprocal connections between the somatosensory cortex and the thalamic Po nucleus in the rat. *Brain Research 477,* 169–174.

Oades, R. D., & Halliday, G. M. (1987). Ventral tegmental (A10) system: Neurobiology, 1. Anatomy and connectivity. *Brain Research Review, 12,* 117–165.

Oakley, D. A. (1979). Cerebral cortex and adaptive behavior. In D. A. Oakley & H. C. Plotkin (Eds.), *Brain, behavior and evolution* (pp. 154–188). London: Methuen.

Oakley, D. A. (1981). Performance of decorticated rats in a two-choice visual discrimination apparatus. *Behavioral Brain Research 3,* 55–69.

Oakley, D. A. (1983a). Learning capacity outside neocortex in animals and man: Implications for therapy after brain-injury. In G. C. L. Davey (Ed.), *Animal models of human behavior* (pp. 247–266). New York: John Wiley.

Oakley, D. A. (1983b). The varieties of memory: A phylogenetic approach. In A. Mayes (Ed.), *Memory in animals and humans* (pp. 20–82). Workingham: Van Nostrand Reinhold.

O'Hearn, E., & Molliver, M. E. (1984). Organization of raphe-cortical projections in rat: A quantitative retrograde study. *Brain Research Bulletin, 13,* 709–726.

Ojemann, G. A. (1982). Interrelationships in the localization of language, memory, and motor mechanisms in human cortex and thalamus. In R. A. Thompson & J. R. Green (Eds.), *New perspectives in cerebral localization* (pp. 157–175). New York: Raven Press.

O'Keefe, J., & Nadel, L. (1978). *The hippocampus as a cognitive map.* Oxford: Clarendon Press.

Olds, J., Disterhoft, J. F., Segal, M., *et al.* (1972). Learning centers of rat brain mapped by measuring latencies of conditioned unit responses. *Journal of Neurophysiology, 35,* 202–219.

Olds, J., Mink, W. D., & Best, P. J. (1969). Single unit patterns during anticipatory behavior. *Electroencephalography and Clinical Neurophysiology, 26,* 144–158.

Orbach, J. (1959). "Functions" of striate cortex and the problem of mass action. *Psychological Bulletin, 56,* 271–292.

Oyanagi, K., Nakashima, S., Ikuta, F., & Homma, Y. (1986). An autopsy case of dementia and parkinsonism with severe degeneration exclusively in the substantia nigra. *Acta Neuropathologica, 70,* 90–192.

Papez, J. W. (1937). A proposed mechanism of emotion. *Archives of Neurology and Psychiatry, 38,* 725–743.

Pasquier, D. A., & Villar, M. J. (1982). Specific serotonergic projections to the lateral geniculate body from the lateral cell groups of the dorsal raphe nucleus. *Brain Research, 249,* 142–146.

Passingham, R. E. (1987). From where does the motor cortex get its instructions? In S. P. Wise (Ed.), *Higher brain functions* (pp. 7–97). New York: Wiley.

Pavlidis, G. T., & Fisher, D. F. (1986). *Dyslexia: Its neuropsychology and treatment.* New York: Wiley.

Paxinos, G. (1985a). *The rat nervous system. Vol. 1. Forebrain and midbrain.* New York: Academic Press.

Paxinos, G. (1985b). *The rat nervous system. Vol. 2. Hindbrain and spinal cord.* New York: Academic Press.

Paxinos, G., & Watson, C. (1982). *The rat brain in stereotaxic coordinates.* New York: Academic Press.

Penfield, W. (1952). Epileptic automatism and the centrencephalic system. *Research Publications for the Association of Nervous and Mental Disorders, 30,* 513–528.

Penfield, W. (1954a). Mechanisms of voluntary movement. *Brain, 77,* 1–17.

Penfield, W. (1954b). Studies of the cerebral cortex of man: A review and an interpretation. In J. F. Delafresnaye (Ed.), *Brain mechanisms and consciousness* (pp. 284–304). Springfield, IL: Charles C Thomas.

Penfield, W. (1958). Centrencephalic integrating system. *Brain, 81,* 231–234.

Penfield, W. (1963). Speech and perception. *Rehabilitation Monograph, 23,* 23–48.

Penfield, W. (1966). Speech, perception and the cortex. In J. C. Eccles (Ed.), *Brain and conscious experience* (pp. 217–237). New York: Springer–Verlag.

Penfield, W. (1975a). The mind and the brain. In F. G. Worden, J. P. Swazey, & G. Adelman (Eds.), *The neurosciences: Paths of discovery* (pp. 437–454). Cambridge, MA: MIT Press.

Penfield, W. (1975b). *The mystery of the mind.* Princeton, NJ: Princeton University Press.

Penfield, W., & Jasper, J. (1954). *Epilepsy and the functional anatomy of the human brain.* Boston: Little, Brown.

Perry, T. L., Norman, M. G., Yong, V. W., *et al.* (1985). Hallervorden–Spatz disease: Cysteine accumulation and cysteine dioxygenase deficiency in the globus pallidus. *Annals of Neurology, 18,* 482–489.

Phillips, A. G., & Carr, G. D. (1987). Cognition and the basal ganglia: A possible substrate for procedural knowledge. *Canadian Journal of Neurological Science, 14,* 381–385.

Piaget, J. (1952). *The origins of intelligence in children, 2nd edition.* New York: International Universities Press.

Piaget, J. (1970). Piaget's theory. In P. H. Mussen (Ed.), *Carmichael's handbook of developmental psychology, 3rd edition* (pp. 703–732). New York: Wiley.

Piercy, M. (1969). Neurological aspects of intelligence. In P. J. Vinken & G. W. Bruyn (Eds.), *Handbook of clinical neurology, Vol. 3* (pp. 296–315). Amsterdam: North-Holland.

Pillon, B., Dubois, B., Lhermitte, F., & Agid, Y. (1986). Heterogeneity of cognitive impairment in progressive supranuclear palsy, Parkinson's disease, and Alzheimer's disease. *Neurology, 36,* 1179–1185.

Porteus, S. D. (1950). *The Porteus maze test and intelligence.* Palo Alto, CA: Pacific Books.

Posner, M. I., Peterson, S. E., Fox, P. T., & Raichle, M. E. (1988). Localization of cognitive operations in the human brain. *Science, 240,* 1627–1631.

Potegal, M. (1972). The caudate nucleus egocentric localization system. *Acta Neurobiologiae Experimentalis, 32,* 479–494.
Pribram, K. H. (1984). Brain systems and cognitive learning processes. In H. L. Roitblat, T. G. Bever, & H. S. Terrace (Eds.), *Animal cognition* (pp. 627–656). Hillsdale, NJ: Lawrence Erlbaum.
Purpura, D. P. (1979). Pathobiology of cortical neurons in metabolic and unclassified amentias. In R. Katzman (Ed.), *Congenital and acquired cognitive disorders* (pp. 43–68). New York: Raven Press.
Ray, C. L., Mirsky, A. F., & Pragay, E. B. (1982). Functional analysis of attention-related unit activity in the reticular formation of the monkey. *Experimental Neurology, 77,* 544–562.
Reader, A. V. (1974). *Machine, brain and mind.* Manchester: A. V. Reader.
Revusky, S. (1985). The general process approach to animal learning. In T. D. Johnston & A. T. Pietrewics (Eds.), *Issues in the ecological study of learning* (pp. 401–432). Hillsdale, NJ: Lawrence Erlbaum.
Ribak, C. E., & Peters, A. (1975). An autoradiographic study of the projections from the lateral geniculate body of the rat. *Brain Research, 92,* 341–368.
Robertson, R. T., Kaitz, S. S., & Robards, M. J. (1980). A subcortical pathway links sensory and limbic systems of the forebrain. *Neuroscience Letters, 17,* 161–165.
Roger, M., & Cadusseau, J. (1984). Afferent connections of the nucleus posterior thalami in the rat, with some evolutionary and functional considerations. *Journal für Hirnforshung, 25,* 473–485.
Roland, P. E. (1985). Cortical organization of voluntary behavior in man. *Human Neurobiology, 4,* 155–167.
Roland, P. E., Meyer, E., Shibasaki, *et al.* (1982). Regional cerebral blood flow changes in cortex and basal ganglia during voluntary movements in normal human volunteers. *Journal of Neurophysiology, 48,* 467–480.
Runnels, L. K., Thompson, R., & Runnels, P. (1968). Near-perfect runs as a learning criterion. *Journal of Mathematical Psychology, 5,* 362–368.
Saavedra, M. A., Pinto-Hamuy, T., & Oberti, C. (1965). Auditory avoidance behavior after extensive and restricted neocortical lesions in the rat. *Journal of Comparative and Physiological Psychology, 60,* 41–45.
Saint-Cyr, J. A., Taylor, A. E., & Lang, A. E. (1988). Procedural learning and neostriatal dysfunction in man. *Brain, 111,* 941–959.
Schwartz, M. F., & Schwartz, B. (1984). In defence of organology. *Cognitive Neuropsychology, 1,* 25–42.
Scott, B. S., Becker, L. E., & Petit, T. L. (1983). Neurobiology of Down's syndrome. *Progress in Neurobiology, 21,* 199–237.
Scott, S., Caird, F., & Williams, B. (1984). Evidence for an apparent sensory speech disorder in Parkinson's disease. *Journal of Neurology, Neurosurgery and Psychiatry, 47,* 840–843.
Semmes, J. (1967). Manual stereognosis after brain injury. In J. F. Bosma (Ed.), *Symposium on oral sensation and perception* (pp. 137–148). Springfield, IL: Charles C Thomas.

Sergent, J. (1987). A new look at the human split brain. *Brain, 110,* 1375–1392.
Shallice, T. (1982). Specific impairments in planning. *Philosophical Transactions of Royal Society, London, 298,* 199–209.
Shammah-Lagnado, S. J., Negrao, N., & Ricardo, J. A. (1985). Afferent connections of the zona incerta: A horseradish peroxidase study in the rat. *Neuroscience, 15,* 109–134.
Shammah-Lagnado, S. J., Negrao, N., Silva, B. A., & Ricardo, J. A. (1987). Afferent connections of the nuclei reticularis pontis oralis and caudalis: A horseradish peroxidase study in the rat. *Neuroscience, 20,* 961–981.
Shaw, C-M. (1987). Correlates of mental retardation and structural changes of the brain. *Brain and Development, 9,* 1–8.
Sherman, B. S., Hoehler, F. K., & Buerger, A. A. (1982). Instrumental avoidance conditioning of increased leg lowering in the spinal rat. *Physiology & Behavior, 25,* 123–128.
Shoenfeld, T. A., & Hamilton, L. W. (1977). Secondary brain changes following lesions. *Physiology and Behavior, 18,* 951–967.
Simons, D., Puretz, S., & Finger, S. (1975). Effects of serial lesions of somatosensory cortex and further neodecortication on tactile retention in rats. *Experimental Brain Research, 23,* 353–365.
Snider, R. S., & Maiti, A. (1976). Cerebellar contributions to the Papez circuit. *Journal of Neuroscience Research, 2,* 133–146.
Snider, R. S., Maiti, A., & Snider, S. R. (1976). Cerebellar pathways to ventral midbrain and nigra. *Experimental Neurology, 53,* 714–728.
Spearman, C. (1923). *The nature of "intelligence" and the principles of cognition.* London: Macmillan.
Spearman, C. (1927). *The abilities of man.* New York: Macmillan.
Sperry, R. W. (1955). On the neural basis of the conditioned response. *British Journal of Animal Behavior, 3,* 41–44.
Sperry, R. W. (1961). Cerebral organization and behavior. *Science, 133,* 1749–1757.
Sperry, R. W. (1982). Forebrain commissurotomy and conscious awareness. In J. Orbach (Ed.), *Neuropsychology after Lashley* (pp. 497–522). Hillsdale, NJ: Lawrence Erlbaum.
Spiliotis, P. H., & Thompson, R. (1973). The "manipulative response memory system" in the white rat. *Physiological Psychology, 1,* 101–114.
Squire, L. R. (1987). *Memory and brain.* New York: Oxford University Press.
Squire, L. R., & Butters, N. (1984). *Neuropsychology of memory.* New York: Guilford Press.
Steele, J. C., Richardson, J. C., & Olszewski, J. (1964). Progressive supranuclear palsy. *Archives of Neurology, 10,* 333–359.
Stein, D. G., Rosen, J. J., & Butters, N. (1974). *Plasticity and recovery of function in the central nervous system.* New York: Academic Press.
Stepien, I., Stepien, L., & Konorski, J. (1961). The effects of unilateral and bilateral ablations of sensorimotor cortex on instrumental (type II) alimentary conditioned reflexes. *Acta Biologiae Experimentalis, 21,* 121–140.

Stern, Y., & Mayeux, R. (1986). Intellectual impairment in Parkinson's disease. *Advances in Neurology, 45,* 405–408.

Sternberg, R. J. (1985). *Beyond I.Q.: A triarchic theory of human intelligence.* New York: Cambridge University Press.

Stokes, L. D., & Thompson, R. (1970). Combined damage to the medial cerebral peduncle and anterior hypothalamus and escape behavior in the rat. *Journal of Comparative and Physiological Psychology, 71,* 303–310.

Storozhuk, V. M., Ivanova, S. F., & Tal'nov, A. N. (1985). Role of the midbrain periaqueductal gray matter in conditioned reflexes. *Neurophysiology, 16,* 320–332.

Stuss, D. T., & Benson, D. F. (1986). *The frontal lobes.* New York: Raven Press.

Swanson, L. W., & Kohler, C. (1986). Anatomical evidence for direct projections from the entorhinal area to the entire cortical mantle in the rat. *Journal of Neuroscience, 6,* 3010–3023.

Switzer, R. C., de Olmos, J., & Heimer, L. (1985). Olfactory system. In G. Paxinos (Ed.), *The rat nervous system. Vol. 1, Forebrain and midbrain* (pp. 1–36). New York: Academic Press.

Taghzouti, K., Simon, H., Herve, G., *et al.* (1988). Behavioral deficits induced by an electrolytic lesion of the rat ventral mesencephalic tegmentum are corrected by a superimposed lesion of the dorsal noradrenergic system. *Brain Research, 440,* 172–176.

Takada, M., & Hattori, T. (1987). Organization of ventral tegmental areas cells projecting to the occipital cortex and forebrain in the rat. *Brain Research, 418,* 27–33.

Taylor, A. E., Saint-Cyr, J. A., & Lang, A. E. (1986). Frontal lobe dysfunction in Parkinson's disease. *Brain, 109,* 845–883.

Thomas, R. K. (1970). Mass action and equipotentiality: A reanalysis of Lashley's retention data. *Psychological Reports, 27,* 899–902.

Thomas, R. K., & Weir, V. K. (1975). The effects of lesions in the frontal or posterior association cortex of rats on maze III. *Physiological Psychology, 3,* 210–214.

Thompson, R. (1959). Learning in rats with extensive neocortical damage. *Science, 129,* 1223–1224.

Thompson, R. (1965). Centrencephalic theory and interhemispheric transfer of visual habits. *Psychological Review, 72,* 385–398.

Thompson, R. (1969). Localization of the "visual memory system" in the white rat. *Journal of Comparative and Physiological Psychology Monograph, 69,* Part 2, 1–29.

Thompson, R. (1974). Localization of the "maze memory system" in the white rat. *Physiological Psychology, 2,* 1–17.

Thompson, R. (1976). Stereotaxic mapping of brainstem areas critical for memory of visual discrimination habits in the rat. *Physiological Psychology, 4,* 1–10.

Thompson, R. (1978a). *A behavioral atlas of the rat brain.* New York: Oxford University Press.

Thompson, R. (1978b). Localization of a "passive avoidance memory system" in the white rat. *Physiological Psychology, 6,* 263–274.
Thompson, R. (1979). Dissociation of a visual discrimination task into incentive, location and response habits. *Physiology & Behavior, 23,* 63–68.
Thompson, R. (1982a). Functional organization of the rat brain. In J. Orbach (Ed.), *Neuropsychology after Lashley* (pp. 207–228). Hillsdale, NJ: Lawrence Erlbaum.
Thompson, R. (1982b). Evidence that the occipital cortex also functions in place learning in rats. In C. Ajmone Marsan (Ed.), *Neural plasticity and memory formation* (pp. 453–463). New York: Raven Press.
Thompson, R. (1983). Brain systems and long-term memory. *Behavioral and Neural Biology, 37,* 1–45.
Thompson, R. (1984). Nonspecific neural mechanisms involved in learning and memory in the rat. In L. R. Squire & N. Butters (Eds.), *Neuropsychology of memory* (pp. 408–416). New York: Guilford Press.
Thompson, R., & Bachman, M. K. (1979a). The occipito–striate pathway and visual discrimination performance in the white rat. *Bulletin of the Psychonomic Society, 14,* 433–434.
Thompson, R., & Bachman, M. K. (1979b). Zona incerta: A link between the visual cortical sensory system and the brainstem motor system. *Physiological Psychology, 7,* 251–253.
Thompson, R., & Peddy, C. P. (1979). The lateral pedunculo–nigral area and visually guided behavior. *Physiology and Behavior, 23,* 1049–1055.
Thompson, R., & Spiliotis, P. H. (1981). Subcortical lesions and retention of a brightness discrimination in the rat: Appetitive vs. aversive motivation. *Physiological Psychology, 9,* 63–67.
Thompson, R., & Yu, J. (1987). The neuroanatomy of learning and memory in the rat. In N. W. Milgram, C. M. MacLeod, & T. L. Petit (Eds.), *Neuroplasticity, learning and memory* (pp. 231–263). New York: Alan R. Liss.
Thompson, R., Arabie, G. J., & Sisk, G. B. (1976). Localization of the "inclined plane discrimination memory system" in the white rat. *Physiological Psychology, 4,* 311–324.
Thompson, R., Gates, C. E., & Gross, S. A. (1979). Thalamic regions critical for retention of skilled movements in the rat. *Physiological Psychology, 7,* 7–21.
Thompson, R., Hale, D. B., & Bernard, B. A. (1980). Brain mechanisms concerned with left–right differentiation in the white rat. *Physiological Psychology, 8,* 309–319.
Thompson, R., Gallardo, K., & Yu, J. (1984a). Thalamic mechanisms underlying acquisition of latch-box problems in the white rat. *Acta Neurobiologiae Experimentalis, 44,* 105–120.
Thompson, R., Gallardo, K., & Yu, J. (1984b). Cortical mechanisms underlying acquisition of latch-box problems in the white rat. *Physiology and Behavior, 32,* 809–817.
Thompson, R., Harmon, D., & Yu, J. (1984). Detour problem-solving behav-

ior in rats with neocortical and hippocampal lesions: A study of response flexibility. *Physiological Psychology, 12,* 116–124.

Thompson, R., Harmon, D., & Yu, J. (1985). Deficits in response inhibition and attention in rats rendered mentally retarded by early subcortical brain damage. *Developmental Psychobiology, 18,* 483–499.

Thompson, R., Huestis, P. W., Crinella, F. M., & Yu, J. (1986). The neuroanatomy of mental retardation in the white rat. *Neuroscience and Biobehavioral Review, 10,* 317–338.

Thompson, R., Huestis, P. W., Crinella, F. M., & Yu, J. (1987). Further lesion studies on the neuroanatomy of mental retardation in the white rat. *Neuroscience and Biobehavioral Review, 11,* 415–440.

Thompson, R., Huestis, P. W., & Yu, J. (1987). Motor learning: Nonspecific subcortical mechanisms in rats. *Archives of Physical Medicine and Rehabilitation, 68,* 419–422.

Thompson, R., Bjelajac, V. M., Huestis, P. W., *et al.* (1989a). Puzzle-box learning impairments in young rats with lesions to the "general learning system." *Psychobiology, 17, 77–88.*

Thompson, R., Bjelajac, V. M., Huestis, P. W., et al. (1989b). Inhibitory deficits in rats rendered "mentally retarded" by early brain damage. *Psychobiology, 17,* 61–76.

Thompson, R., Bjelajac, V. M., Fukui, S., *et al.* (1989). Failure to transfer a digging response to a detour problem in young rats with lesions to the "general learning system." *Physiology & Behavior, 45,* 1235–1241.

Thompson, R. F. (1986). The neurobiology of learning and memory. *Science, 233,* 941–947.

Thompson, R. F. (1987). Identification of an essential memory trace circuit in the mammalian brain. In N. W. Milgram, C. M. MacLeod, & T. L. Petit (Eds.), *Neuroplasticity, learning, and memory* (pp. 151–172). New York: Alan R. Liss.

Thompson, R. F., Clark, G. A., Donegan, N. H., *et al.* (1984). Neuronal substrates of learning and memory: A "multiple-trace" view. In G. Lynch, J. L. McGaugh, & N. M. Weinberger (Eds.), *Neurobiology of learning and memory* (pp. 137–164). New York: Guilford Press.

Thomson, G. H. (1951). *The factorial analysis of human ability, 5th edition.* Boston: Houghton Mifflin.

Thorndike, R. L. (1935). Organization of behavior in the albino rat. *Genetic Psychology Monographs, 17,* 1–70.

Thurstone, L. L. (1947). *Multiple factor analysis.* Chicago: University of Chicago Press.

Torack, R. M., & Morris, J. C. (1988). The association of ventral tegmental area histopathology with adult dementia. *Archives of Neurology, 45,* 497–501.

Towbin, A. (1969). Mental retardation due to germinal matrix infarction. *Science, 164,* 151–161.

Tracey, D. J. (1985). Somatosensory system. In G. Paxinos (Ed.), *The rat ner-*

vous system. Vol. 2, Hindbrain and spinal cord (pp. 129–152). New York: Academic Press.
Trimble, M. R., & Cummings, J. L. (1981). Neuropsychiatric disturbances following brainstem lesions. *British Journal of Psychiatry, 138,* 56–59.
Tryon, R. C. (1931). Studies of individual differences in maze ability. III. The community of function between two maze abilities. *Journal of Comparative Psychology, 12,* 95–115.
Tryon, R. C. (1959). Domain sampling formulation of cluster and factor analysis. *Psychometrika, 24,* 113–135.
Tryon, R. C., & Bailey, D. E. (1970). *Cluster analysis.* Toronto: McGraw–Hill.
Ueno, T., & Takahata, N. (1978). Chronic brainstem encephalitis with mental symptoms and ataxia. *Journal of Neurology, Neurosurgery and Psychiatry, 41,* 516–524.
Van der Kooy, D., & Kolb, B. (1985). Non-cholinergic globus pallidus cells that project to the cortex but not to the subthalamic nucleus in the rat. *Neuroscience Letters, 57,* 113–118.
Vanderwolf, C. H. (1964). Effect of combined medial thalamic and septal lesions on active-avoidance behavior. *Journal of Comparative and Physiological Psychology, 58,* 31–37.
Viaud, F. (1960). *Intelligence: Its evolution and forms.* London: Hutchinson.
Vilkki, J. (1988). Problem solving deficits after focal cerebral lesions. *Cortex, 24,* 119–127.
Villar, M. J., Vitale, M. L., Hokfelt, T., & Verhofstad, A. A. J. (1988). Dorsal raphe serotonergic branching neurons projecting both to the lateral geniculate body and superior colliculus: A combined retrograde tracing–immunohistochemical study in the rat. *Journal of Comparative Neurology, 277,* 126–140.
Voneida, T. J. (1967). The effect of pyramidal lesions on the performance of a conditioned avoidance response in cats. *Experimental Neurology, 19,* 483–493.
Vonsattel, J-P., Myers, R. H., Stevens, T. J., *et al.* (1985). Neuropathological classification of Huntington's disease. *Journal of Neuropathology and Experimental Neurology, 44,* 559–577.
Voogd, J., Gerritts, N. M., & Marani, E. (1985). Cerebellum. In G. Paxinos (Ed.), *The rat nervous system. Vol. 2. Hindbrain and spinal cord* (pp. 251–292). New York: Academic Press.
Wallesch, C. W., Kornhuber, H. H., Brunner, R. J., *et al.* (1983). Lesions of the basal ganglia, thalamus, and deep white matter: Differential effects on language functions. *Brain and Language, 20,* 286–304.
Walshe, F. M. R. (1957). The brain-stem conceived as the "highest level" of function in the nervous system; with particular reference to the "automatic apparatus" of Carpenter (1850) and to the "centrencephalic integrating system" of Penfield. *Brain, 80,* 510–539.
Warren, J. M. (1961). Individual differences in discrimination learning by cats. *Journal of Genetic Psychology, 98,* 89–93.

Watson, J. B. (1907). Kinesthetic and organic sensations: Their role in the reactions of the white rat to the maze. *Psychological Monograph, 8,* No. 2, 1–100.

Webster, W. G. (1973). Assumptions, conceptualizations, and the search for the functions of the brain. *Physiological Psychology, 1,* 346–350.

Weiskrantz, L. (1987). Neuroanatomy of memory and amnesia. A case for multiple memory systems. *Human Neurobiology, 6,* 93–105.

Wenk, G. L., Cribbs, B., & McCall, L. (1984). Nucleus basalis magnocellularis: Optimal coordinates for selective reduction of choline acetyltransferase in frontal neocortex by ibotenic acid injections. *Experimental Brain Research, 56,* 335–340.

Wertheimer, M. (1959). *Productive thinking.* New York: Harper & Row.

Whitehouse, P. J. (1986). The concept of subcortical and cortical dementia: Another look. *Annals of Neurology, 19,* 1–6.

Winer, J., & Lubar, J. F. (1976). Alternation behavior of cats with medial visual cortex ablation. *Physiology & Behavior, 17,* 635–643.

Wise, S. P., & Jones, E. G. (1977). Cells of origin and terminal distribution of descending projections of the rat somatic sensory cortex. *Journal of Comparative Neurology, 175,* 129–158.

Wyss, J. M., Swanson, L. W., & Cowan, W. M. (1979). A study of subcortical afferents to the hippocampal formation in the rat. *Neuroscience, 4,* 463–476.

Zaidel, E. (1981). Hemispheric intelligence: The case of the Raven Progressive Matrices. In M. P. Friedman, J. P. Das, & N. O'Conner (Eds.), *Intelligence and learning* (pp. 531–552). New York: Plenum Press.

Zangwill, O. L. (1964). Lashley's concept of cerebral mass action. *Proceedings of the Royal Society of Medicine, 57,* 914–917.

Zeaman, D., & House, B. J. (1967). The relation of IQ and learning. In R. M. Gagne (Ed.), *Learning and individual differences* (pp. 192–217). Columbus, OH: Charles E. Merrill.

Zemlan, F. P., Leonard, C. M., & Pfaff, D. W. (1978). Ascending tracts of the lateral columns of the rat spinal cord: A study using the silver impregnation and horseradish peroxidase techniques. *Experimental Neurology, 62,* 298–334.

Zigler, E., & Butterfield, E. C. (1968). Motivational aspects of changes in IQ test performance of culturally deprived nursery school children. *Child Development, 39,* 1–14.

Zoladek, L., & Roberts, W. A. (1978). The sensory basis of spatial memory in the rat. *Animal Learning & Behavior, 6,* 77–81.

Zubek, J. P. (1951). Studies in somesthesis: I. Role of somesthetic cortex in roughness discrimination in the rat. *Journal of Comparative and Physiological Psychology, 44,* 339–353.

Index

Lesions and corresponding illustrations are listed on pages 53–55; learning tasks and corresponding maps are listed on page 89.

GPSR Compliance

The European Union's (EU) General Product Safety Regulation (GPSR) is a set of rules that requires consumer products to be safe and our obligations to ensure this.

If you have any concerns about our products, you can contact us on ProductSafety@springernature.com

In case Publisher is established outside the EU, the EU authorized representative is:

Springer Nature Customer Service Center GmbH
Europaplatz 3
69115 Heidelberg, Germany

Batch number: 10761588

Printed by Printforce, the Netherlands